Intelligent Perception and Information Processing

This book series Intelligent Perception and Information Processing is grounded in artificial intelligence and signal and information processing theory, focusing on key technologies such as communication, vision, and location for signal and information perception, processing, transmission, and storage. It is committed to solving key and difficult problems in wireless perception, machine vision, information and coding, image processing, multi-source information fusion, environment configurable intelligent transmission, cross-media transmission, knowledge graph and other application areas, and promoting the development of intelligent perception and signal processing.

Hongyang Gu · Yao Ding · Xiaogang Yang ·
Ruitao Lu · Lei Pu · Siming Han

Automated Machine Learning for Person Re-Identification

Springer

Hongyang Gu
Xi'an Research Institute of High Technology
Xi'an, Shaanxi, China

Xiaogang Yang
Xi'an Research Institute of High Technology
Xi'an, Shaanxi, China

Lei Pu
Xi'an Research Institute of High Technology
Xi'an, Shaanxi, China

Yao Ding
Xi'an Research Institute of High Technology
Xi'an, Shaanxi, China

Ruitao Lu
Xi'an Research Institute of High Technology
Xi'an, Shaanxi, China

Siming Han
Xi'an Research Institute of High Technology
Xi'an, Shaanxi, China

ISSN 3059-3808 ISSN 3059-3816 (electronic)
Intelligent Perception and Information Processing
ISBN 978-981-95-3432-6 ISBN 978-981-95-3433-3 (eBook)
https://doi.org/10.1007/978-981-95-3433-3

This Springer imprint is published by the registered company Springer Nature Singapore Pte Ltd.
The registered company address is: 152 Beach Road, #21-01/04 Gateway East, Singapore 189721, Singapore

Preface

Person re-identification (ReID) has emerged as a critical technology in intelligent video surveillance systems, enabling the tracking and identification of individuals across nonoverlapping camera views. Despite significant advances in deep learning, designing effective ReID systems remains challenging due to variations in pose, illumination, occlusion, and camera settings. Traditional methods often rely heavily on manual design of data augmentation strategies, network architectures, and loss functions—a process that is both time-consuming and suboptimal.

This book explores the integration of **Automated Machine Learning (AutoML)** into person re-identification, aiming to reduce human intervention and enhance model performance through automated optimization of key components. We focus on three core aspects: **automated data augmentation**, **neural architecture search**, and **loss function search**, each tailored to the unique characteristics of ReID tasks.

The book is structured into seven chapters. Chapter 1 introduces the background, significance, and current challenges of ReID, along with relevant datasets and evaluation metrics. Chapters 2 through 6 present novel AutoML methods developed specifically for ReID, including:

- Differentiable erasing-based data augmentation (Chap. 2).
- Auto-scale single-branch and multi-scale multi-branch architecture search (Chaps. 3 and 4).
- Dynamic and static margin-based softmax loss function search (Chaps. 5 and 6).

Chapter 7 concludes the book and suggests directions for future research.

This work is intended for researchers, engineers, and graduate students in computer vision, machine learning, and video surveillance. We hope this book will serve as a valuable resource for those interested in leveraging AutoML to advance the state of the art in person re-identification.

We would like to thank all collaborators and reviewers who contributed to this book. Special thanks to our families and institutions for their unwavering support.

Xi'an, China
August 2025

Hongyang Gu
Yao Ding
Xiaogang Yang
Ruitao Lu
Lei Pu
Siming Han

Contents

Chapter 1
Introduction

1.1 Background and Significance of the Research Project

With the rapid development of economies and societies worldwide, Intelligent Video Surveillance (IVS) has become a crucial technology and is being widely deployed in various critical locations, such as schools, shopping malls, hospitals, train stations, airports, strategic sites, and other important venues (Dufour 2012). Person Re-identification (ReID) is a core and key technology within the entire IVS system, aiming to search for and identify specific individuals across different camera scenes (Gheissari et al. 2006). The overall framework of the ReID system is illustrated in Fig. 1.1. The process begins with acquiring video sequence images from the camera network in a given area, followed by the use of person detection techniques to accurately locate individuals within the video sequences. This results in a database of images of individuals captured across multiple cameras. Finally, the system performs ReID by matching the query image of a person with all positive samples in the database that correspond to the same identity label.

With the continuous improvement of large-scale surveillance systems, person re-identification technology has played a significant role in scenarios such as tracking criminal suspects and locating missing elderly people and children. In intelligent video surveillance, face recognition technology also holds an important position. However, due to the typical high placement and wide coverage of surveillance cameras, it is often difficult to obtain high-resolution facial images, which poses a significant challenge for identity recognition using face recognition technology. In more challenging scenarios, suspects often obscure their faces with masks, hoods, sunglasses, and hats, rendering face recognition technology completely ineffective. In contrast, person re-identification can accurately identify individuals by extracting features from the entire body. Therefore, as an important complement to face recognition technology, person re-identification holds considerable application value. Moreover, current person re-identification technology involves numerous research

H. Gu et al., *Automated Machine Learning for Person Re-Identification*,
Intelligent Perception and Information Processing,
https://doi.org/10.1007/978-981-95-3433-3_1

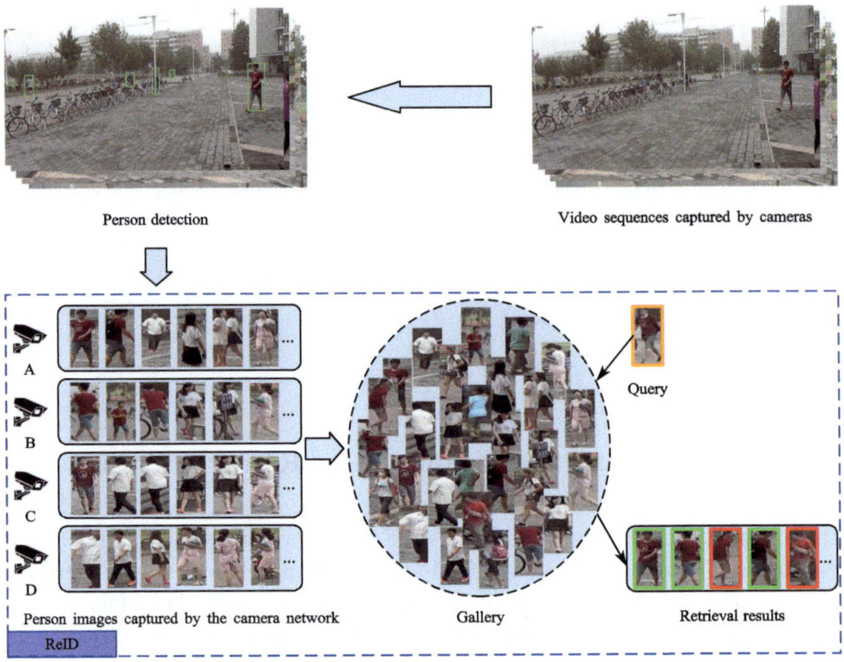

Fig. 1.1 The overall framework of the ReID system

techniques and fields, and in-depth studies of it also have theoretical significance for other related areas.

The core challenge in person re-identification lies in extracting robust features from person images. Early traditional research in person re-identification focused on manually extracting visual features from person images and designing corresponding distance measurement algorithms. However, due to various influencing factors such as posture, background, camera parameters, lighting, and detector performance, the traditional approach of separately designing feature extraction and distance measurement proves inadequate for current complex application scenarios (Zheng et al. 2016; Asperti et al. 2025). In recent years, with the proliferation of training data and the advancement of computing power, deep learning techniques based on artificial neural networks have gradually become the mainstream method in person re-identification. Compared with traditional methods, deep learning-based person re-identification approaches leverage deep learning's automatic feature learning capabilities to integrate feature extraction and distance measurement learning into a single framework, thereby facilitating the acquisition of more abstract and robust features. However, the current focus of most deep learning-based person re-identification methods remains on the design of network architectures (Ye et al. 2022). These architectures are primarily based on designs originally intended for fundamental tasks such as image classification, object detection, and object segmentation. Such architectures are often suboptimal for the specific task of person re-identification. Moreover, for deep neural models, data augmentation strategies and loss functions also play a crucial role. Nevertheless, current deep learning-based person re-identification methods rarely prioritize these two aspects. Therefore, it is imperative to design more effective data augmentation strategies, more refined network architectures, and more reasonable loss functions that align with the unique characteristics of the person re-identification task.

Currently, numerous effective techniques have been developed for deep learning-based person re-identification (He et al. 2023). These include data augmentation techniques like Random Erasing (RE) (Zhong et al. 2020); network architectures such as the Part-based Convolutional Baseline (PCB) (Sun et al. 2018), Omni-Scale Network (OSNet) (Zhou et al. 2022), and Multiple Granularities Network (MGN) (Wang et al. 2018); and loss functions like Cross-Entropy Loss, Triplet Loss (Schroff et al. 2015), and CircleLoss (Sun et al. 2020). While these techniques demonstrate good performance, a significant gap still exists between their current capabilities and practical deployment requirements. Specifically, in some novel application scenarios, certain techniques may exhibit limitations, often necessitating further fine-tuning with expert knowledge to achieve satisfactory results. This process is cumbersome, time-consuming, and does not consistently yield optimal outcomes.

Automated Machine Learning (AutoML) is an emerging research direction that integrates automation with machine learning, aiming to automate the machine learning process to reduce or even completely eliminate the need for human expertise. In recent years, the focus of AutoML research has shifted from traditional machine learning models to deep learning models (Zöller and Huber 2021; Baratchi et al. 2024; He et al. 2021), leading to several popular research directions, such as Automated Data Augmentation (ADA) (Cubuk et al. 2019; Ho et al. 2019; Wang et al. 2021; Li et al. 2025), Neural Architecture Search (NAS) (Zoph and Le 2017; Real et al. 2017; Zoph et al. 2018; Chu et al. 2020; Ye et al. 2022; Liwei et al. 2024; Avval et al. 2025), and Loss Function Search (LFS) (Wu et al. 2018; Li et al. 2019; Liu and Lai 2020; Zhao et al. 2021; Huang et al. 2022; Li et al. 2022). In these three research directions, AutoML methods have already demonstrated superior performance compared to manually designed methods, attracting increasing attention from researchers. However, the current focus of AutoML research remains primarily on fundamental tasks such as image classification, object detection, and object segmentation, with comparatively limited application in person re-identification, which is inherently a metric learning task. Therefore, conducting research on AutoML-based technologies, tailored to the specific characteristics of the person re-identification task, is a highly meaningful topic. This research not only helps to further improve the performance of deep learning-based person re-identification models but also significantly contributes to the development of AutoML-related theories.

This book presents several rational and effective AutoML methods for data augmentation, network architecture, and loss functions, specifically designed for deep learning-based person re-identification models and informed by their unique characteristics.

1.2 Research Status

1.2.1 Automated Machine Learning

Currently, machine learning, as a data-driven and powerful modeling method, has become one of the most cutting-edge research areas in artificial intelligence (Zhou 2021). Looking at the modeling process of machine learning, it can generally be divided into four stages: data preparation, feature engineering, model selection and training, and model evaluation and optimization (He et al. 2021). Each of these four stages directly affects the overall performance of the model. Therefore, the entire modeling process requires intervention and guidance of experts. However, this adjustment process often consumes a significant amount of time and effort and is not very user-friendly for non-experts. In recent years, with the rise of deep learning, deep neural networks (DNNs) have demonstrated excellent performance in areas such as computer vision, natural language processing, and speech recognition due to their powerful automatic feature learning capabilities. However, the complexity of deep neural network models has reached an unprecedented level compared to traditional models. For example, the ResNet-50 network (He et al. 2016) has as many as 26M parameters, and even experts find it difficult to design a sophisticated network architecture based on ResNet-50 for new application scenarios in a short period of time.

Automated machine learning (AutoML) is a technology that aims to automate machine learning processes. It not only allows non-experts to efficiently build high-performing machine learning models but also provides potential inspiration for experts in their subsequent design work. Based on the types of models to which AutoML is applied, it can be divided into traditional AutoML methods and deep learning-based AutoML methods. Traditional AutoML methods mainly focus on traditional machine learning models (such as support vector machines (Cortes and Vapnik 1995), logistic regression (Tolles and Meurer 2016), and k-nearest neighbors (Fix 1985) and, in combination with the machine learning modeling process, primarily conduct localized automated searches in two aspects: automatic model selection (Max Kanter and Veeramachaneni 2015; Christ et al. 2018; Kaul et al. 2017) and automatic feature engineering (Thornton et al. 2013; Feurer et al. 2022). AutoML-Zero (Real et al. 2020) is Google's work on automating the entire machine learning process from scratch. It can not only reproduce algorithms manually designed by humans over the years but also discover novel and effective algorithms previously unknown to researchers. These results validate the possibility of automatically searching for more novel and effective machine learning algorithms in the future for some challenging problems, further highlighting the importance and prospects of AutoML.

Deep learning-based AutoML methods focus on deep neural networks. Compared to traditional machine learning models, deep neural networks can learn good features directly from raw data. Therefore, the research emphasis of deep learning-based AutoML methods is on the network architecture and training process of deep neural networks themselves. Since the training process of deep neural networks often takes a longer time, deep learning-based AutoML methods need to pay more attention to the efficient training and evaluation of models than traditional AutoML methods. Generally speaking, the core of deep learning-based AutoML methods is to solve the following bi-level optimization problem (Dempe 2002):

$$
\begin{aligned}
s^{\star} &= \underset{s \in \mathcal{S}}{\arg\max}\, \mathcal{P}_V\left(w^{\star}(s), s\right), \\
\text{s.t.} w^{\star}(s) &= \underset{w \in \mathcal{W}}{\arg\min}\, \mathcal{L}_T(w, s),
\end{aligned}
\tag{1.1}
$$

where s represents a specific configuration in the search space $\mathcal{S}$, w denotes the weight parameters of the deep neural network model, and $\mathcal{W}$ indicates the space set where the weight parameters reside. $\mathcal{L}_T(w, s)$ represents the loss function used for the deep neural network weight parameters w on the current configuration s and training data $\mathcal{D}_{trn}$. $w^*(s)$ denotes the optimal solution for the inner-level problem, i.e., the deep neural network weight parameters corresponding to the minimum $\mathcal{L}_T(w, s)$. $\mathcal{P}_V(w^*(s), s)$ represents the performance of the deep neural network on the validation data $\mathcal{D}_{val}$ when the current configuration and deep neural network weight parameters are s and $w^*(s)$, respectively. s^* denotes the optimal solution for the outer-level problem, i.e., the desired optimal target. According to different search objectives, corresponding search spaces $\mathcal{S}$ can be designed. Currently, there is corresponding work in data augmentation (Cubuk et al. 2019; Ho et al. 2019; Wang et al. 2021; Li et al. 2025), network architectures (Zoph and Le 2017; Real et al. 2017; Zoph et al. 2018; Chu et al. 2020; Ye et al. 2022; Liwei et al. 2024; Avval et al. 2025), loss functions (Wu et al. 2018; Li et al. 2019; Liu and Lai 2020; Zhao et al. 2021; Huang et al. 2022; Li et al. 2022), regularization (Pham and Le 2021), activation functions (Sun et al. 2024), and optimization (Bello et al. 2017). Combining the research content of this book, the focus will be on introducing automated data augmentation, neural architecture search, and loss function search.

1.2.1.1 Automated Data Augmentation

Data augmentation is a technique for increasing the diversity of training data, playing a crucial role in preventing model overfitting and enhancing model generalization. Currently, data augmentation techniques are widely applied in fields such as vision, natural language, and speech (Shorten and Khoshgoftaar 2019). In line with the content of this book, this chapter focuses on data augmentation for visual images. The key point of automated data augmentation research lies in data augmentation based on transformations, which involves using some predefined rules to generate new data from the original data. Common transformations include geometric transformations, color transformations, filtering operations, and image mixing (Zhong et al. 2020; Inoue 2018; Yun et al. 2019; Chen et al. 2020), as shown in Fig. 1.2. Data augmentation operations based on transformations have a core design principle: the augmented images can be considered to have unchanged image labels, but this principle is difficult to ensure. Taking the task of person re-identification as an

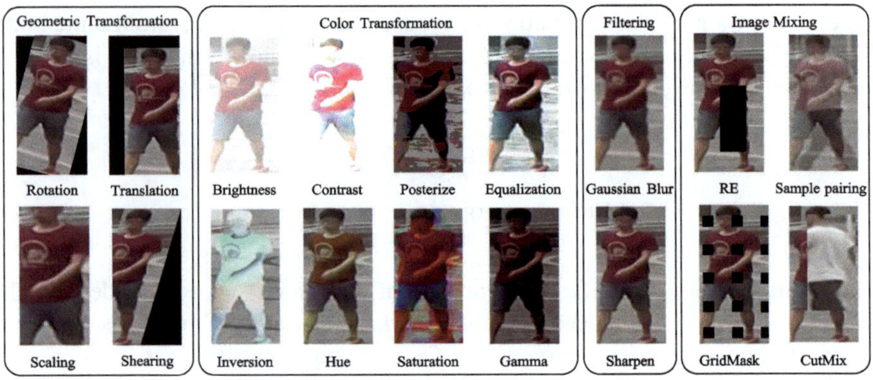

Fig. 1.2 Common transformation-based data augmentation operations

example, a moderate degree of geometric transformation is in line with this principle, but when the geometric transformation is too intense, it will inevitably lead to the loss of key features. Additionally, inappropriate operations are also detrimental to the task of person re-identification, such as the "inversion" operation, which can cause drastic changes in the color characteristics of a person. Therefore, an effective data augmentation strategy should be a proper augmentation operation and appropriate augmentation intensity under the principle of unchanged image labels after augmentation. Designing a strategy entirely by hand is time-consuming and labor-intensive, while automated data augmentation can specifically search for effective data augmentation strategies for a given task, ensuring that even in the absence of expert knowledge, an effective augmentation strategy can be obtained.

As a pioneering work in automated data augmentation, AA (Cubuk et al. 2019) employs Reinforcement Learning (RL) to search for data augmentation strategies. Specifically, AA uses Recurrent Neural Networks (RNNs) as controllers to generate augmentation strategies and evaluates their effectiveness through training deep models from scratch. The evaluation results are then used to update the controller's parameters to generate better augmentation strategies. Although the strategies found by AA outperform hand-designed strategies in image classification tasks, the method of strategy evaluation brings significant computational complexity. For example, searching on the ImageNet dataset requires approximately 15000 GPU hours to obtain a good strategy. Such a huge computational resource demand is difficult for most researchers to meet, hence the focus of subsequent research is on how to ensure effective augmentation strategies while improving search efficiency.

Inspired by PBT (Jaderberg et al. 2017), PBA (Ho et al. 2019), OHL-Auto-Aug (Lin et al. 2019), and Adv.AA (Yang et al. 2022), automated data augmentation is transformed into an online hyperparameter optimization problem, that is, the search for dynamic augmentation strategies. Under multi-GPU parallelism, these three methods, respectively, use evolutionary algorithms, reinforcement learning, and adversarial learning to complete the search in just one normal training time. Among them, Adv.AA's performance is the most superior and has surpassed AA's performance. Gontijo-Lopes et al. (2021) analyzed that AA's search space is inherently friendly to datasets such as CIFAR and ImageNet, hence Adv.AA, which uses adversarial learning, can achieve better performance. Similarly, RandAugment (2020), which benefits from the friendly design of the search space, can achieve performance comparable to AA by searching only for the number and intensity of operations. Fast AA (Lim et al. 2019), based on the principle of density matching and using Bayesian optimization, reduces the search time on the ImageNet dataset to 450 GPU hours and achieves performance comparable to AA. Faster AA (Hataya et al. 2020) and DADA (Li et al. 2020) use the Gumbel-Softmax technique (Jang et al. 2017) to approximate the non-differentiable augmentation process as a differentiable process, then use differentiable search algorithms to reduce the search time to 2.3 GPU and 1.3 GPU hours, respectively, and also achieve performance comparable to AA. AWS (Tian et al. 2020) proposed a weight-sharing network for data augmentation strategies, and based on this, even using AA's reinforcement learning method can improve the search efficiency by about 39 times, and the strategies found can surpass the performance of AA and Adv.AA. MCMC-Aug (Tian et al. 2021) further continuousizes the search space, decomposing the data augmentation strategy into three sequentially executed differentiable modules: color adjustment, image filtering, and image transformation. Combined with the weight-sharing network and differentiable search proposed by AWS, MCMC-Aug's search efficiency can be higher than DADA, and the strategies found can maintain performance comparable to AWS.

In addition to being applied to image classification tasks such as CIFAR and ImageNet, automated data augmentation has also achieved good performance in object detection and point cloud classification fields (Zoph et al. 2020; Li et al. 2020a, 2025; Mumuni and Mumuni 2025). Applying automated data augmentation in new fields not only reduces the dependence on expert knowledge but also helps improve the performance levels in these fields.

1.2.1.2 Neural Architecture Search

Deep neural networks can automatically learn high-level and abstract features, which is inseparable from their excellent network architectures. Different network architectures have a crucial impact on different tasks. Therefore, many research efforts focus on the design of neural network architectures (He et al. 2016). However, designing an excellent neural network architecture requires certain expert knowledge and experience. For non-experts, it is generally difficult to adjust the network architecture appropriately according to their actual application needs in a short time. Neural architecture search can automatically design a network architecture with better performance for the current task without human intervention, consuming certain computational resources.

As pioneering work in neural architecture search, NAS-RL (Zoph and Le 2017), MetaQNN (Baker et al. 2017), Large-scale Evolution (Real et al. 2017), and GeNet (Xie and Yuille 2017) use reinforcement learning or evolutionary algorithms for neural architecture search. However, they have some shortcomings, which can be summarized as follows (Dosovitskiy et al. 2021):

(1) Using a global search space, the representation of network architectures is overly redundant and has poor scalability.
(2) It is necessary to train the network architecture from scratch to evaluate its performance, leading to high search costs.
(3) The search strategy is discrete, which is not conducive to combining with efficient differentiable search algorithms.

Subsequent neural architecture search work mainly improves from the above three aspects. The problem of the global search space is mainly solved by using a modular search space, that is, only the local modules in the network architecture are searched, and then the entire network architecture is obtained by repeatedly stacking the modules. This modular search space can not only represent the entire network architecture more compactly but also achieve better scalability by increasing or decreasing the number of modules. NASNet (Zoph et al. 2018a) is a representative work using a modular search space. It mainly searches for two types of modules—normal modules and downsampling modules. Normal modules and downsampling modules are modules that extract features under the premise of whether the resolution of the feature map changes. The entire network architecture is generated by repeatedly stacking multiple normal modules before each downsampling module. In addition, NASNet no longer searches for the specific configuration of each operation but searches for the connection relationships of operations in the module, thus greatly reducing the search space. It also uses the same reinforcement learning method as NAS-RL. Experimental results show that the network architecture searched, whether in terms of the number of parameters or classification accuracy, has significantly outperformed NAS-RL. This shows the importance of a good search space for neural architecture search, and subsequent work has focused on this modular search space.

To quickly evaluate the performance of architectures, ENAS (Pham et al. 2018) proposed a weight-sharing mechanism. It treats each architecture to be evaluated as a subgraph in a directed acyclic hypergraph and shares the weights of the architectures sampled each time. ENAS also uses a recurrent neural network as a controller to sample subgraphs on the hypergraph. Only the weights of the subgraphs sampled are activated, and the remaining weights do not participate in backpropagation training. The update of the controller mainly uses reinforcement learning. In this way, the controller parameters and hypergraph weights are optimized alternately. Compared to previous work, ENAS has greatly improved search efficiency. On a single GPU, it can complete the search on the CIFAR-10 dataset in less than a day.

Compared to reinforcement learning and evolutionary algorithms, differentiable optimization is faster. DARTS (Liu et al. 2018) first pioneered differentiable architecture search. It uses a module-based search space like NASNet and uses a differentiable framework for efficient search. Specifically, DARTS treats each module as a directed acyclic graph composed of N sequentially executed nodes, where each node represents a feature $x^{(i)}$, and the directed edge from node i to node j represents

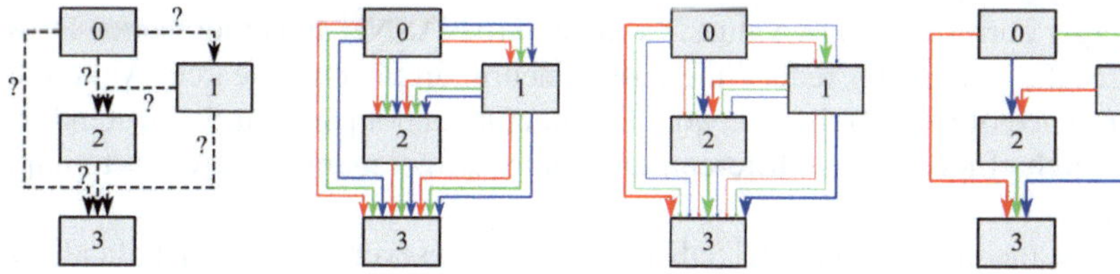

(a) Operations to be determined (b) Continuous representation (c) After optimization (d) Discretization

Fig. 1.3 The process of DARTS optimization

an operation $o^{(i,j)}$ (such as 3×3 convolution and pooling operations). In a discrete manner, the feature of each node can be expressed as:

$$x^{(j)} = \sum_{i<j} o^{(i,j)}(x^{(i)}). \tag{1.2}$$

DARTS uses Softmax to relax Eq. (1.2) into a continuous form, thus obtaining a mixed output for all operations $\mathcal{O}$:

$$\overline{o}^{(i,j)}(x) = \sum_{o \in \mathcal{O}} \frac{\exp(\alpha_o^{(i,j)})}{\sum_{o' \in \mathcal{O}} \exp(\alpha_{o'}^{(i,j)})} o(x), \tag{1.3}$$

where $\mathcal{O}$ represents the set of all candidate operations, and $\alpha_o^{(i,j)}$ represents the weight of operation o on edge $e^{(i,j)}$. Under this relaxation, neural architecture search becomes the optimization of continuous architecture parameters $\alpha = \{\alpha^{(i,j)}\}$. Once the search is completed, the operation with the maximum weight on edge $e^{(i,j)}$ will be selected as $o^{(i,j)}$:

$$o^{(i,j)} = \arg\max_{o \in \mathcal{O}} \alpha_o^{(i,j)}. \tag{1.4}$$

DARTS transforms discrete operation selection into continuous operation selection. Furthermore, it can transform the standard bi-level optimization problem of automatic machine learning in Eq. (1.1) into:

$$\begin{aligned} \alpha^\star =& \arg\min_{\alpha \in \mathcal{A}} \mathcal{L}_V \left(w^\star(\alpha), \alpha\right), \\ \text{s.t. } w^\star(\alpha) =& \arg\min_{w \in \mathcal{W}} \mathcal{L}_T(w, \alpha), \end{aligned} \tag{1.5}$$

where $\mathcal{A}$ represents the parameter space of architecture parameters α. By introducing the loss function $\mathcal{L}_V(w^\star(\alpha), \alpha)$ on the validation dataset $\mathcal{D}_{val}$ in the outer optimization problem, DARTS can use gradients to update weight parameters and architecture parameters. Figure 1.3 shows the DARTS optimization process with four nodes, where lines of different colors represent different operations, and thicker lines represent larger corresponding α values. Compared to NASNet and other works, DARTS, like ENAS, can greatly reduce the cost of neural architecture search. Although the performance of the searched network architecture is not particularly outstanding, its high search efficiency and differentiable nature attract many researchers to improve DARTS.

Firstly, the process of DARTS discretization has several issues. Specifically, after the DARTS supernetwork converges, the probabilities obtained for some operations through the Softmax function are not exactly zero. Directly selecting the operation corresponding to the maximum weight value through Eq. (1.4) will cause a significant gap between the supernetwork and the subnetwork. SNAS (Xie et al. 2019) remodels neural architecture search from a probabilistic perspective, theoretically ensuring the consistency between the supernetwork and the subnetwork. DARTS+ (Liang et al. 2019) uses early stopping to prevent excessive skip connections in the subnetwork, thereby improving the performance of the subnetwork. Fair DARTS (Chu et al. 2020) replaces the original Softmax in DARTS with Sigmoid and combines it with a 0–1 regularization term to alleviate this problem.

Secondly, the high consumption of DARTS on GPUs leads to a gap in network architecture between search and retraining, that is, a smaller architecture is used during search. P-DARTS (Chen et al. 2021) ensures that the gap between the network architecture during search and retraining will not be too large through a progressive approach. GDAS (Dong and Yang 2019) uses the Gumbel-Softmax trick (Jang et al. 2017) to sample sub-networks and only backpropagates the parameters of the subnetwork, greatly reducing GPU consumption. PC-DARTS (Xu et al. 2020) uses channel sampling to alleviate this problem.

In addition to image classification tasks, neural architecture search has wide applications in other areas such as object detection and segmentation (Chen et al. 2019; Wu et al. 2024; Qin et al. 2023). Furthermore, in addition to searching for Convolutional Neural Network (CNN) architectures, there are a lot of interests in searching for Vision Transformer (ViT) (Dosovitskiy et al. 2021) and Mamba (Gu and Dao 2023) architectures (Zhou and Zhu 2024; Zhao et al. 2025; Qin et al. 2024).

1.2.1.3 Loss Function Search

Loss functions have always been an important research subject in machine learning. They define a measure of the loss that balances the model's performance and optimize the model's parameters by minimizing this loss. Deep neural networks, as models with tens of thousands of parameters, often use Stochastic Gradient Descent (SGD) to update parameters. Therefore, the loss functions of deep neural networks must have differentiable properties. Generally, target metrics are often non-differentiable (such as classification accuracy in classification problems and mean average precision in retrieval problems), which requires the design of surrogate loss functions that can approximate these metrics and are differentiable for the optimization of deep neural networks. Designing good loss functions often requires a lot of expert knowledge,

Table 1.1 Classification of loss function search

Loss function type	Search scope	Examples
Static	Global	GLO (Gonzalez and Miikkulainen 2020), CSE-Autoloss (Liu et al. 2021), AutoLoss-Zero (Li et al. 2022), TaylorGLO (Gonzalez and Miikkulainen 2021)
	Local	Auto Seg-Loss (Li et al. 2021), Para-AP (Tao et al. 2021)
Dynamic	Global	LS (Grabocka et al. 2019), EPBT (Zhi Liang et al. 2020)
	Local	L2T-DLF (Wu et al. 2018), AutoLoss (Xu et al. 2018), SLF (Liu and Lai 2020), AutoLoss-R (Zhao et al. 2021)
		AM-LFS (Li et al. 2019), Search-Softmax (Wang et al. 2020)

and hyperparameter optimization of them is also quite laborious. Therefore, using loss function search for automatic loss function design is very meaningful.

According to the form of the loss functions searched, they can be divided into dynamic loss function search and static loss function search. The former's form is continuously adjusted and changed during training, while the latter's loss function remains unchanged throughout the entire training process. Looking at the search scope of the loss function, it can also be divided into local search and global search. Global search aims to discover the entire mathematical structure of a loss function from scratch, while local search only searches a part of the loss function. Table 1.1 shows the specific methods according to the above classification.

L2T-DLF (Wu et al. 2018) and LS (Grabocka et al. 2019) both use neural networks to perform local dynamic approximation of target loss and global dynamic approximation of target metrics, respectively. However, this approach introduces some prior knowledge when optimizing the outer neural network, making the results likely to be suboptimal. GLO (Gonzalez and Miikkulainen 2020) pioneered the use of computational graphs to represent loss functions. Specifically, it represents the loss function as a computational graph composed of basic mathematical operations and uses genetic programming and CMA-ES optimization strategies to optimize the structure and weights of the computational graph. While this decoupled optimization of structure and weights may lead to a performance gap, using computational graphs to represent loss functions does help discover novel and effective losses. CSE-Autoloss (Liu et al. 2021) and AutoLoss-Zero (Li et al. 2022) are two works that continue the representation of computational graphs. They propose a series of acceleration strategies based on evolutionary algorithms, and from the search results, both have achieved better performance than manually designed ones. TaylorGLO (Gonzalez and Miikkulainen 2021) innovatively uses Taylor expansion to represent loss functions and uses CMA-ES to optimize the weights. Within the same search time, TaylorGLO achieves better results than GLO, but the forms of loss functions it can represent are relatively limited. Based on the representation of TaylorGLO, EPBT (Zhi Liang et al. 2020) uses an algorithm similar to PBT (Jaderberg et al. 2017) for dynamic loss function search. Compared to static loss function search, dynamic search methods can adjust the form of the loss function according to the current training situation. AutoLoss (Xu et al. 2018), SLF (Liu and Lai 2020), and AutoLoss-R (Zhao et al. 2021) are methods that dynamically adjust the weight ratios of multiple known loss functions. These methods utilize more prior knowledge, making it difficult to find new forms of loss functions. In addition, some work conducts local searches on the framework of existing loss functions, such as AM-LFS (Li et al. 2019) and Search-Softmax (Wang et al. 2020), which are based on the Margin-based Softmax (MS) loss function:

$$\mathcal{L}_{MS} = -\log\left(\frac{\exp(s \cdot t(\cos\theta_y))}{\exp(s \cdot t(\cos\theta_y)) + \sum_{k \neq y} \exp(s \cdot \cos\theta_k)}\right), \tag{1.6}$$

where $\cos\theta_k = \mathbf{w}_k^T \mathbf{x}/(\|\mathbf{w}_k\|\|\mathbf{x}\|)$. Different $t(x)$ can represent different Margin-based Softmax loss functions, such as ArcFace (Deng et al. 2022) and CosFace (Wang et al. 2018). AM-LFS and Search-Softmax both use reinforcement learning to dynamically search for $t(x)$. Specifically, AM-LFS represents $t(x)$ as a piecewise linear function and optimizes the slope and intercept of each segment, while Search-Softmax only searches for m in $t(x) = x - m$. This piecewise function representation is also used in Auto Seg-Loss (Li et al. 2021) and Para-AP (Tao et al. 2021), both of which use piecewise functions to substitute for non-differentiable parts of the target metric and have achieved good performance.

Loss function search is more challenging compared to automated data augmentation and neural architecture search, and there is still much room for research in terms of the representation of the search space and the design of search algorithms.

1.2.2 Person Re-identification

Since the proposal of person re-identification in 2006 Gheissari et al. (2006), it has attracted increasing attention from researchers due to its significant practical and theoretical value. Depending on the application scenarios of person re-identification, different research problems can be identified (Zheng et al. 2016; Ye et al. 2022). From the perspective of data types, it can be divided into image-based person re-identification tasks and video-based person re-identification tasks (Ming et al. 2022). In terms of data modality, it can be divided into single-modality person re-identification tasks and multi-modality person re-identification tasks (Zheng et al. 2022). This book focuses on the most fundamental task in person re-identification–image-based, closed-world (Closed-World) person re-identification. The task is formally defined as follows. Given a surveillance system, the initial dataset $\mathcal{D}_0 = \{(x_i, y_i)|_{i=1}^N\}$ is obtained by manually or using detection algorithms to extract bounding boxes from the video sequences of all cameras, where $x_i \in \mathbb{R}^{H \times W \times 3}$ represents the RGB image of the person, and $y_i \in \{1, 2, \ldots, C\}$ represents the identity label of the person. The dataset contains C persons with different identity labels. The initial dataset $\mathcal{D}_0$ is then divided into the training dataset $\mathcal{D}_{trn} = \{(x_i^{(trn)}, y_i^{(trn)})|_{i=1}^{N_{trn}}\}$ and the testing dataset $\mathcal{D}_{tst} = \{(x_i^{(tst)}, y_i^{(tst)})|_{i=1}^{N_{tst}}\}$, with the following relationships:

$$\begin{cases} N_{trn} + N_{tst} = N \\ \{y_i^{(trn)}|_{i=1}^{N_{trn}}\} \cap \{y_i^{(tst)}|_{i=1}^{N_{tst}}\} = \emptyset \\ \{y_i^{(trn)}|_{i=1}^{N_{trn}}\} \cup \{y_i^{(tst)}|_{i=1}^{N_{tst}}\} = \{1, 2, \ldots, C\} \end{cases}. \tag{1.7}$$

Further, the testing dataset $\mathcal{D}_{tst}$ is divided into the query set $Q = \{(x_i^{(Q)}, y_i^{(Q)})|_{i=1}^{N_Q}\}$ and the gallery set $\mathcal{G} = \{(x_i^{(\mathcal{G})}, y_i^{(\mathcal{G})})|_{i=1}^{N_\mathcal{G}}\}$, with the following relationships:

$$\begin{cases} N_Q + N_\mathcal{G} = N_{tst} \\ \{y_i^{(Q)}|_{i=1}^{N_Q}\} \subseteq \{y_i^{(\mathcal{G})}|_{i=1}^{N_\mathcal{G}}\} = \{y_i^{(tst)}|_{i=1}^{N_{tst}}\} \end{cases}. \tag{1.8}$$

After the dataset is divided, the model can be trained using the training dataset $\mathcal{D}_{trn}$, and based on the well-trained model, the similarity $\text{sim}(x^{(Q)}, x_i^{(\mathcal{G})})$ between the query image $x^{(Q)}$ and each image $x_i^{(\mathcal{G})}$ in the gallery set $\mathcal{G}$ can be obtained. The most ideal retrieval result is that the similarity of all positive samples in the gallery set $\mathcal{G}$ with the same identity label $y^{(Q)}$ as the query image $x^{(Q)}$ is greater than that of the negative samples:

$$\min_{y_i^{(\mathcal{G})} = y^{(Q)}} \text{sim}\left(x^{(Q)}, x_i^{(\mathcal{G})}\right) > \max_{y_i^{(\mathcal{G})} \neq y^{(Q)}} \text{sim}\left(x^{(Q)}, x_i^{(\mathcal{G})}\right). \tag{1.9}$$

According to the above formal definition, it can be seen that there is an essential difference between the person re-identification task and the image classification task. The training and testing datasets in image classification tasks have the same categories, while the testing dataset in person re-identification tasks may contain categories different from those in the training dataset. Therefore, simply treating the person re-identification task as a classification problem can only ensure that it can distinguish well among the categories that have appeared in the training dataset, and the distinction of categories that have not been seen in the testing dataset will be significantly insufficient (Zheng et al. 2016).

In the early stages of person re-identification research, the focus was on designing visual features and distance metric algorithms for person images. In the design of person image features, various statistics were used to generate low-level person image features by combining visual features such as color and texture (Gheissari et al. 2006; Gray and Tao 2008; Zhang et al. 2016). Further, mid-level features based on attributes were combined to generate more robust features (Layne et al. 2012; Li et al. 2016). In terms of distance metric algorithms, most early methods were designed based on the Mahalanobis Distance (Köstinger et al. 2012; Yang et al. 2016). Additionally, subspace-based methods were also a focus of research (Dong et al. 2018), and of

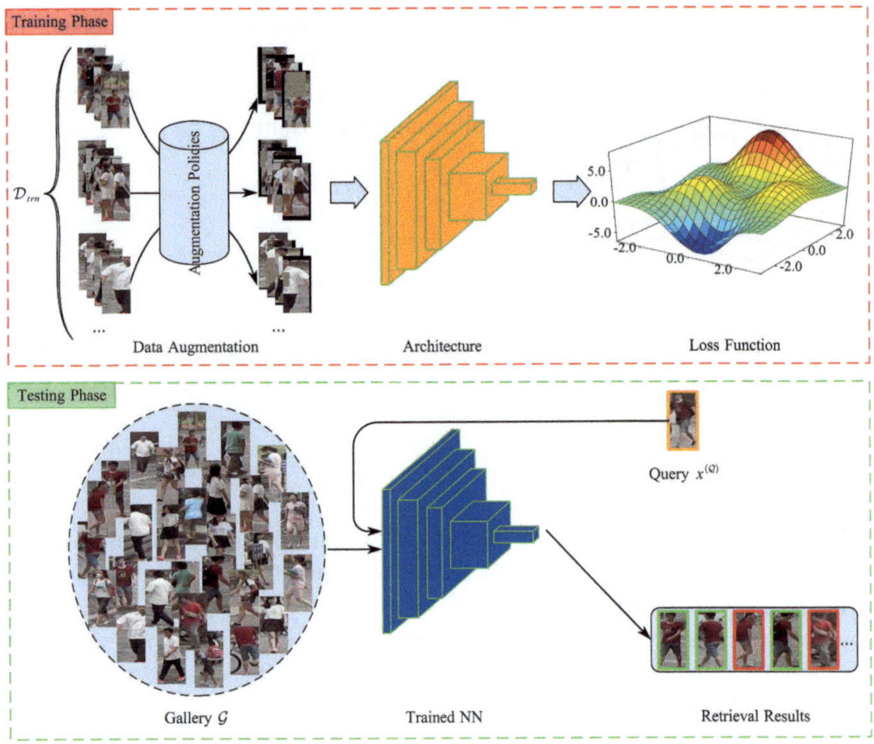

Fig. 1.4 The overall process of deep learning-based person re-identification methods

course, methods using SVM (Cortes and Vapnik 1995) or Boosting (Schapire 2002) also achieved good results.

With the breakthrough development of deep learning technology, significant progress has been made in the fields of image classification, object detection, and segmentation. Research on person re-identification tasks centered around deep learning technology has also increased, achieving many remarkable accomplishments. Currently, deep learning-based person re-identification methods have become the mainstream paradigm in the field of person re-identification, and this book also focuses on systematically introducing the current research status of these methods. Under the setting of image-based, closed-world person re-identification tasks, the overall process of deep learning-based methods is shown in Fig. 1.4. During the training phase, effective data augmentation strategies must first be designed to augment the training dataset $\mathcal{D}_{trn}$. Then, an excellent network architecture is designed to extract features from the augmented data. Finally, a reasonable loss function is used to optimize the network to a convergence point with good generalization performance. During the testing phase, the well-trained deep neural network is used for feature extraction and similarity calculation of the gallery set $\mathcal{G}$ and the query image $x^{(Q)}$, and the retrieval results are obtained based on the similarity. From the entire process, the core factors affecting model performance lie in the design of data augmentation, network architecture, and loss function. The following will provide a systematic overview of the current research status from these three aspects.

1.2.2.1 Data Augmentation

To meet the demand for large sample sizes in deep models, many large-scale datasets have emerged in person re-identification since 2015, such as the Market-1501 dataset (Zheng et al. 2015) containing 1501 individuals and the MSMT17 dataset (Wei et al. 2018) containing 4101 individuals. Manual annotation of large datasets often requires high costs, thus data augmentation has become a convenient and cost-effective technique for expanding existing datasets. In person re-identification, data augmentation can be mainly divided into two categories: data augmentation based on generation and data augmentation based on transformation.

As a pioneering work in using generative adversarial networks (GANs) for person re-identification data augmentation, LSRO (Zheng et al. 2018) first used deep convolutional generative adversarial networks to generate images. Then, it used a semi-supervised learning framework to provide uniform labels for the generated person images while enhancing the original dataset with the generated person image features. Compared to methods that only use the original dataset, LSRO achieves better performance. Chen et al. (2017) used conditional generative adversarial networks to generate new data samples to alleviate the problem of background interference in the original dataset. Since the background generation only considers the background, the effect is better than LSRO. However, considering the limited improvement in background generation, Pose-transfer (Liu et al. 2018) uses the pose of the original person to generate new data by combining the guidance module, thus improving the quality of the generated images and better considering the background. PN-GAN (Qian et al. 2018) uses a generative adversarial network to eliminate the influence of poses in the original dataset and the generated dataset. It uses two separate deep neural networks to learn the features of the original dataset and the generated dataset, and then uses the original features and the eight typical poses of the generated dataset as the final features. This method can also achieve better performance. In addition to eliminating the influence of poses, camera view differences also need to be considered. CamStyle (Zhong et al. 2018) uses CycleGAN to transfer the images of each camera to the style of other cameras, thus reducing the influence of camera view differences to a certain extent. The above methods treat the generation and feature learning of images as two independent stages, and some methods combine the learning of images and features to promote each other. FD-GAN (Ge et al. 2018) uses the distillation method to integrate the generation and feature learning to obtain identity-related and pose-unrelated features, thus eliminating the influence of pose-related information on feature learning. At the same time, FD-GAN avoids the need for input image pose information in the testing phase, further reducing the dependence on pose detection modules. DG-Net (Zheng et al. 2019) is a task-specific model that learns to generate high-quality person images from human appearance and structural features, thus achieving data augmentation. In addition, DG-Net does not require external information input during both training and testing phases, thus completely avoiding the introduction of noise. However, DG-Net does not fully utilize the feature information of persons, which is not conducive to feature learning. IS-GAN (Eom and Ham 2019) fully learns the features related to and unrelated to persons, thus achieving more prominent performance. JoT-GAN (Zhao et al. 2022) further studies this joint training method and introduces a feature separation module in DG-Net, making feature learning more direct. Although data augmentation based on generation can intuitively obtain generated images, its training complexity and performance do not match, often resulting in a counterproductive effect. Therefore, data augmentation based on transformation, due to its simplicity and convenience, has been favored more.

Data augmentation based on transformation uses predefined rules to transform raw data into new data. In person re-identification tasks, the commonly used paradigms are Random Horizontal Flipping (RH), Random Cropping (RC), and Random Erasing (RE) (Zhong et al. 2020). Among them, Random Erasing erases a rectangular area from the original image with a preset probability p, allowing the model to learn more detailed person features. Specifically, first, a rectangular area size S_e is randomly sampled based on the original image area S and the preset minimum and maximum area values S_l, S_h. Then, the aspect ratio r_e of the rectangular area is randomly sampled based on the preset minimum and maximum aspect ratios r_l, r_h. The length and width of the rectangular area can then be determined as $H_e = \sqrt{S_e \times r_e}$ and $W_e = \sqrt{S_e / r_e}$, respectively. Finally, the center coordinates of the rectangular area (x_e, y_e) are determined based on the length and width H, W of the original image. The erased rectangular area generated in this way does not guarantee that it will be within the original image. In practice, there will be N attempts to generate the erased area. If none of these N attempts are within the original image, the original image is output. Although Random Erasing is simple, it greatly enhances the robustness of the person re-identification model, so many works consider Random Erasing as a key technique for performance improvement. In addition to the random generation method of Random Erasing, AOS (Huang et al. 2018) uses a pre-trained model to obtain adversarial erased areas and then re-trains the model with these adversarial erased areas. Although AOS can achieve good performance, its training process is more complex and the effect is not prominent. In most cases, using Random Erasing is sufficient. FastReID (He et al. 2023) introduced the augmentation strategy searched by AA into person re-identification. Since this strategy is mainly aimed at ImageNet image classification tasks and contains many operations that are not friendly to person re-identification, such as color space transformations, it actually reduces the model's performance after using this strategy.

Currently, there is still much room for research on data augmentation based on transformation in person re-identification tasks. To avoid hindering person re-identification tasks at the data level, it is necessary to conduct relevant research.

1.2.2.2 Network Architecture

Currently, most research on deep learning-based person re-identification methods focuses on network architecture design. Corresponding research works exist from classical Convolutional Neural Networks to the popular Vision Transformer (Dosovitskiy et al. 2021). The network architectures studied in this book refer to those that can obtain features corresponding to each image through the network during the testing phase, and then utilize simple Euclidean distance or cosine similarity to obtain the similarity between two images. This approach is more efficient during the testing phase compared to directly using the network to output the similarity of two images (Ahmed et al. 2015), and thus has become the current mainstream paradigm. According to this approach, existing architectures can be divided into single-branch global feature architecture, single-branch local feature architecture, single-branch global-local feature architecture, multi-branch global feature architecture, and multi-branch global-local feature architecture, as shown in Fig. 1.5. Specifically, a single-branch architecture refers to an architecture with only one branch, where the final features are all derived from the same feature map. In contrast, a multi-branch architecture contains multiple branches, and its final features are derived from different feature maps. Global feature architectures extract features through global feature aggregation operations, while local feature architectures extract features through local feature aggregation operations. Global-local feature architectures incorporate both global and local feature aggregation operations. Figure 1.6 illustrates the specific classification of current deep learning-based person re-identification methods

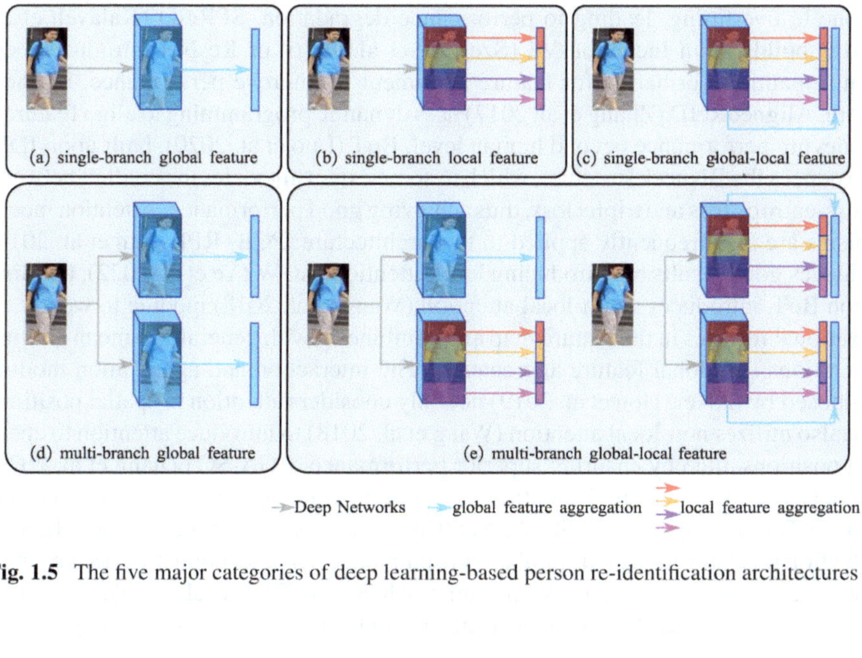

Fig. 1.5 The five major categories of deep learning-based person re-identification architectures

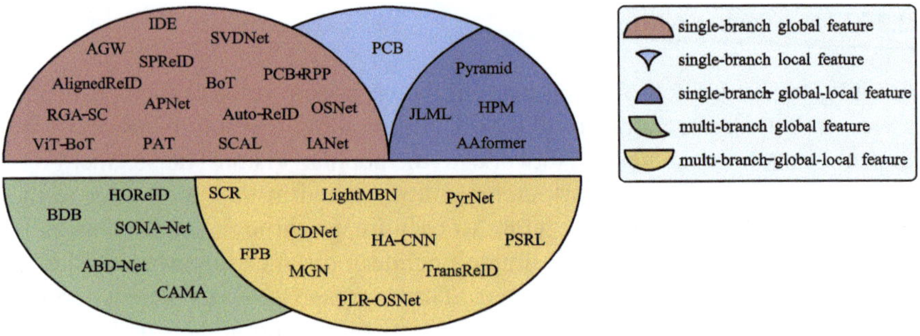

Fig. 1.6 Methods for deep learning-based person re-identification based on network architecture classification

according to the aforementioned five major architectural categories. Detailed introductions to representative works in each category will be provided below.

The single-branch global feature architecture is the simplest and most widely studied architecture. The IDE (Zheng et al. 2016) network uses ResNet-50 as the backbone, aggregating global features through global uniform pooling, and performing identity classification with the simplest cross-entropy loss. Currently, the IDE network is widely used as a benchmark network architecture for person re-identification (Zheng et al. 2018; Zhong et al. 2017). SVDnet (Sun et al. 2017), building upon IDE, decorrelates the last layer of the network using singular value decomposition and efficiently achieves this decorrelation through a proposed tension-relaxation iterative strategy. However, single-branch global feature architectures are prone to overfitting, leading to performance degradation. SPReID (Kalayeh et al. 2018) builds upon Inception-V3 (Szegedy et al. 2016) or ResNet introduces person semantic information for feature alignment to improve performance. Furthermore, AlignedReID (Zhang et al. 2017) uses dynamic programming to align features, achieving performance beyond human level. BoT (Luo et al. 2020), built upon IDE, introduces the BNneck structure, which to some extent alleviates the conflict between cross-entropy loss and triplet loss, thus achieving good performance. Attention mechanisms are also frequently applied to this architecture; PCB+RPP (Sun et al. 2018) achieves good results by introducing local attention. AGW (Ye et al. 2022), building upon BoT, introduces a non-local attention (Wang et al. 2018) module to weight all positional features in the feature map and combines it with generalized mean pooling operations for global feature aggregation. The interaction and aggregation module proposed by IANet (Hou et al. 2019) not only considers attention to spatial positions but also utilizes non-local attention (Wang et al. 2018) to introduce attention to channel positions, thereby ensuring superior performance. RGA-SC (Zhang et al. 2020) models spatial and channel attention in a more flexible way, achieving good performance. Building upon RGA-SC, APNet (Chen et al. 2021) proposes a pyramid-based attention mechanism, which simulates the human visual perception process through a splitting-attending-merging-stacking approach. SCAL (Chen et al. 2019) treats attention as an agent in reinforcement learning and optimizes attention by training a critic, achieving good results with this novel capability. Furthermore, there is a renewed emphasis on directly employing fully attention-based Vision Transformer models. ViT-BoT (He et al. 2021) directly incorporates the BNneck proposed by BoT into ViT (Dosovitskiy et al. 2021), also achieving good performance. PAT (Li et al. 2021) starts from the Transformer decoder and proposes local prototype embedding for feature alignment. In addition to using Inception, ResNet, and ViT as backbone networks, OSNet (Zhou et al. 2022) manually designs a specific network structure for the person re-identification task. This architecture naturally integrates multi-scale information and achieves good performance while being lightweight. Auto-ReID (Quan et al. 2019) automatically designs a single-branch global feature architecture for person re-identification tasks using neural architecture search.

Currently, among single-branch local feature architectures, PCB (Sun et al. 2018) is prominent. It obtains local features through horizontal partitioning. Although simple, this method achieves good performance and has therefore been widely adopted in many subsequent works for local feature extraction. However, because the acquisition of local features inevitably introduces noise to some extent, there are not many works solely relying on local feature architectures. A common practice is to use local features as a supplement to global features (Zhang et al. 2017).

Representative works for single-branch global-local feature architectures include JLML (Li et al. 2017), which extracts local features in the same manner as PCB and global features using global pooling operations. HPM (Fu et al. 2019) and Pyramid (Zheng et al. 2019) have similar architectures, pushing the PCB horizontal partitioning method for local feature extraction to its extreme by obtaining local features containing multi-scale information through horizontal partitioning of different sizes. AAformer (Zhu et al. 2024), based on a Transformer encoder, combines optimal transport theory to propose a multi-head auto-alignment module for local feature extraction.

Multi-branch architectures can utilize multi-level features, thus achieving superior overall performance compared to single-branch architectures. Multi-branch global feature architectures perform global feature extraction on each branch. BDB (Dai et al. 2019), built upon ResNet-50, not only extracts global features using a global branch but also employs a feature dropping branch for more fine-grained global feature extraction. SONA-Net (Bryan et al. 2019) has two branches consistent with BDB, and it also incorporates an attention module into its backbone network. Inspired by class activation maps, CAMA (Yang et al. 2019) proposed an overlapping activation penalty to encourage the class activation maps of different branches to focus on distinct regions, thereby obtaining multiple global features focusing on different information. ABD-Net (Chen et al. 2019), built upon ResNet-50, employs spatial and channel attention similar to non-local attention (Wang et al. 2018) and combines it with orthogonal regularization to constrain the feature maps, thereby ensuring that global features of different degrees are learned across two branches. HOReID (Wang et al. 2021) proposed global hierarchical pooling and local hierarchical pooling to extract multi-level global features, achieving the effect of feature alignment.

The multi-branch global-local feature architecture can integrate global, local, multi-level, and fine-grained features, making it currently the most cutting-edge and, like the single-branch global feature architecture, a widely studied architecture. Pyranet (Martinel et al. 2019) proposed a pyramid module that integrates multi-scale structural features. MGN (Wang et al. 2018) is a further extension of PCB and Pyramid on branches, capable of extracting multi-branch fine-grained features, and has become a highly representative architecture in person re-identification, with many subsequent works based on this design idea. SCR (Chen et al. 2020), building upon MGN, further proposed extracting local features through channel partitioning. PLR-OSNet (Xie et al. 2020) uses OSNet as its backbone network and introduces non-local attention to extract global and local features separately using two branches. LightMBN (Herzog et al. 2021), also based on OSNet, utilizes three branches to extract global, channel-based, and local features separately. HA-CNN (Li et al. 2018) manually designs a lightweight architecture using Inception modules, incorporating various soft attention mechanisms to extract global and local features. FPB (Zhang et al. 2021), building upon global features, proposed a feature pyramid branch composed of attention mechanisms and cross-orthogonal regularization to extract features from different layers. TransReID (He et al. 2021), built upon ViT-BoT, cleverly integrates camera ID information into the model through additional information embedding and extracts local features using a patch-level module. CDNet (Li et al. 2021) searches for efficient multi-branch architectures for person re-identification tasks within a fusion depth search space. Combined with the proposed FBLNeck, the searched architecture has only 1.8 M parameters. In addition to the above methods for obtaining local features, PSRL (Li et al. 2021) extracts local features using person semantic information and combines person structural relationships with Graph Convolutional Neural Networks to learn structure-based global features.

From the five major architectural categories classified, it can be observed that multi-branch local feature architectures and single-branch local feature architectures are still largely unexplored. As mentioned, relying solely on local features is unreliable. The two currently popular architectures are the single-branch global feature architecture and the multi-branch global-local feature architecture. However, both of these architectures still have some shortcomings, necessitating further in-depth research.

1.2.2.3 Loss Function

The loss function is the crucial core for guiding deep neural networks to optimal points. A well-designed loss function can achieve a multiplier effect, thus making loss functions a key research area in deep learning. Currently, there are three main types of loss functions commonly used in deep learning-based person re-identification methods: classification-based loss functions, metric learning-based loss functions, and hybrid loss functions combining classification and metric learning. Classification-based loss functions treat person re-identification as a classification problem, often employing cross-entropy loss:

$$\mathcal{L}_{ce} = -\frac{1}{n}\sum_{i=1}^{n} \log(p(y_i|x_i)), \tag{1.10}$$

where n denotes the number of training samples in a batch, and $p(y_i|x_i)$ represents the probability that person image x_i is recognized as identity label y_i. The specific form of $p(y_i|x_i)$ can vary, with the most common being standard Softmax (Zheng et al. 2016; Sun et al. 2018, 2017):

$$p_0(y_i|x_i) = \frac{\exp(w_{y_i}^T f_i + b_{y_i})}{\exp(w_{y_i}^T f_i + b_{y_i}) + \sum_{c \neq y_i} \exp(w_c^T f_i + b_c)}, \tag{1.11}$$

where $f_i = \mathcal{M}(x_i) \in \mathbb{R}^d$ denotes the features extracted from person image x_i by neural network $\mathcal{M}$. $w_c \in \mathbb{R}^d$ and $b_c \in \mathbb{R}$, respectively, represent the weight and bias corresponding to identity label c, with a total of C different person identity labels. Furthermore, introducing label smoothing into the standard Softmax cross-entropy loss function can effectively prevent model overfitting (Zheng et al. 2018; Luo et al. 2020). Besides the standard Softmax form, some works have investigated its variants, which can be uniformly expressed as:

$$p_1(y_i|x_i) = \frac{\exp(s \cdot t(\cos\theta_{y_i}))}{\exp(s \cdot t(\cos\theta_{y_i})) + \sum_{c \neq y_i} \exp(s \cdot n(\cos\theta_c))}, \tag{1.12}$$

where $\cos\theta_c = w_c^T f_i/(||w_c||||f_i||)$, and both $t(x)$ and $n(x)$ are functions with a domain of $[-1, 1]$. Table 1.2 lists the corresponding $t(x)$ and $n(x)$ for commonly used different loss functions. SphereReID (Fan et al. 2019) sets $t(x)$ and $n(x)$ simply as $y = x$ and uses a larger s, showing better performance than standard Softmax according to experimental results. CircleLoss (Sun et al. 2020) achieves a more flexible loss function and good performance through carefully designed $t(x)$ and $n(x)$. Furthermore, AM-LFS (Li et al. 2019) models $t(x)$ as an M-segment piecewise linear function:

$$\text{pwl}(x; \{a_i|_{i=0}^{M-1}\}, \{b_i|_{i=0}^{M-1}\} = \begin{cases} a_0 x + b_0, & x \in [-1, -1 + 2/M] \\ a_j x + b_j, & x \in (-1 + 2j/M, -1 + 2(j+1)/M] \end{cases}, \tag{1.13}$$

and uses reinforcement learning to dynamically search for the slopes and intercepts $\{a_i|_{i=0}^{M-1}\}, \{b_i|_{i=0}^{M-1}\}$ of each segment.

Features learned by standard Softmax cross-entropy loss functions only possess separability, whereas person re-identification tasks place more emphasis on feature discriminability (Deng et al. 2022; Wang et al. 2018). The aforementioned Softmax variants are one way to enhance feature discriminability. Metric learning-based loss functions offer another effective approach, with the most fundamental being the binary relationship-based contrastive loss (Rama Varior et al. 2016):

Table 1.2 Variants of softmax loss functions in person re-identification

Loss Function	Design Method	$t(x)$	$n(x)$
SphereReID (Fan et al. 2019)	Manual	x	x
CircleLoss (Sun et al. 2020)	Manual	$[\text{de}(1+m-x)]_+(x-1+m)$	$[\text{de}(m+x)]_+(x-m)$
AM-LFS (Li et al. 2019)	Automatic	$\text{pwl}(x; \{a_i\|_{i=0}^{M-1}\}, \{b_i\|_{i=0}^{M-1}\})$	x

$$\mathcal{L}_{con} = \mathbb{I}_{y_i=y_j}[d^2(f_i, f_j)] + \mathbb{I}_{y_i \neq y_j}\left[\max\left(0, m - d(f_i, f_j)\right)^2\right], \tag{1.14}$$

where $d(f_i, f_j)$ denotes the distance between features f_i and f_j (Euclidean distance, L1 distance, etc.), and $\mathbb{I}_c[\cdot]$ is an indicator function that outputs 1 if a condition c is met and 0 otherwise. m is the margin parameter, aiming to ensure that when $y_i \neq y_j$, the distance $d(f_i, f_j)$ between features f_i and f_j is as large as possible, thereby pushing apart features of different identities. Furthermore, triplet-based contrastive loss (Hermans et al. 2017) also has widespread applications:

$$\mathcal{L}_{tri} = [m + d(f_a, f_p) - d(f_a, f_n)]_+, \tag{1.15}$$

where f_a, f_p, f_n denote the anchor, positive sample, and negative sample features, respectively, with the relationship $y_i = y_j \neq y_k$. The objective of $\mathcal{L}_{tri}$ is to ensure that the distance $d(f_a, f_p)$ between features of the same identity (f_a and f_p) is at least m smaller than the distance $d(f_a, f_n)$ between features of different identities (f_a and f_n). However, using the original triplet loss can lead to the problem of "easy triplets" dominating the entire training process, resulting in insufficient discriminability of learned features. This issue is often alleviated by designing effective mining strategies to select meaningful triplets (Hermans et al. 2017; Shi et al. 2016; Song et al. 2018). Currently, the most widely used approach involves hard sample mining within a batch, where the triplet loss can be expressed as:

$$\mathcal{L}_{hard-tri} = [m + \max_{p \in P(a)} d(f_a, f_p) - \min_{n \in N(a)} d(f_a, f_n)]_+, \tag{1.16}$$

where $P(a)$ denotes the set of all indices in a batch that share the same identity label y_a, and $N(a)$ denotes the set of all indices in a batch that do not share the identity label y_a. This mining strategy has become a commonly used triplet loss function in current person re-identification methods.

Even with the hard sample mining strategy for triplet loss, its convergence speed is still slower than $\mathcal{L}_{ce}$. Hybrid loss functions, which combine classification and metric learning, can both accelerate convergence and improve feature discriminability. Among these, $\mathcal{L}_{ce}$ and $\mathcal{L}_{hard-tri}$ have become common paradigms in person re-identification methods (Luo et al. 2020). Currently, for person re-identification tasks, most works still employ manually designed loss functions, with less exploration of new forms of loss functions.

1.3 Current Challenges

Currently, deep learning-based person re-identification methods have achieved good performance, but these methods are still suboptimal in many aspects, making it difficult to meet the practical application demands in complex, large-scale scenarios of current intelligent video surveillance. Therefore, how to accurately and efficiently design and optimize appropriate data augmentation, network architectures, and loss functions for person re-identification tasks based on actual application requirements remains an unresolved problem. Combining the three core components in the overall process of deep learning-based person re-identification methods, the main challenges that need to be overcome are summarized as follows:

(1) More Effective Data Augmentation. Due to the cross-camera retrieval characteristic of person re-identification tasks, manually annotating large datasets is a very expensive and time-consuming process. However, deep learning-based person re-identification methods still rely on large-scale data to avoid network overfitting. Currently, the most common and direct means of data expansion is data augmentation technology. Among these, transformation-based data augmentation is the simplest and most effective method. It transforms original data into new data according to preset rules, thereby achieving data expansion and preventing model overfitting. Currently, commonly used transformation-based data augmentations primarily focus on operations such as random horizontal flipping, random cropping, and random erasing (Zhong et al. 2020). These augmentation operations are based on manual design, making them suboptimal in many aspects. Furthermore, the search space for AA-based automated data augmentation has, to some extent, been verified as not suitable for person re-identification tasks (He et al. 2023). Therefore, at the data level, current augmentation strategies constrain the performance of the entire person re-identification system. Hence, designing more effective data augmentation tailored to the characteristics of person re-identification tasks is meaningful.

(2) More Sophisticated Network Architectures. In person re-identification tasks, human images undergo significant variations due to factors such as pose, illumination, and camera changes. Currently, most researchers are dedicated to designing sophisticated network architectures to mitigate the impact of these factors and thereby enhance the performance of the entire person re-identification system. Most network architectures are designed with backbones used for image classification tasks, such as ResNet, Inception, and DenseNet. However, person re-identification tasks have their peculiarities, making it often inappropriate to directly apply these architectures for person re-identification. Furthermore, most manually designed architectures often have a large number of parameters and high computational complexity, making them difficult to deploy in resource-constrained environments. Although some researchers have manually designed lightweight architectures (Zhou et al. 2022; Li et al. 2018), achieving both lightweightness and high performance simultaneously through manual design remains extremely challenging. Auto-ReID (Quan et al. 2019) and CDNet (Li et al. 2021), which incorporate neural architecture search techniques, can indeed discover lightweight and high-performance network architectures for person re-identification tasks. However, they often do not fully integrate the specific characteristics of person re-identification tasks and incorporate commonly used multi-scale and other key techniques insufficiently. Therefore, at the network architecture level, it is necessary to design more sophisticated network architectures that integrate key person re-identification techniques.

(3) More Reasonable Loss Functions. Currently, the common loss function paradigms for deep learning-based person re-identification methods are cross-entropy loss and triplet loss, with relatively less research on designing new forms of loss functions. SphereReID (Fan et al. 2019) and CircleLoss (Sun et al. 2020) are two manually designed loss functions that both achieve better performance than the standard Softmax cross-entropy loss. However, due to the inherent limitations of manual design, the performance improvements they can achieve are still limited. Although AM-LFS (Li et al. 2019) achieves better results by dynamically searching for loss functions using reinforcement learning, the method itself has certain limitations that restrict its performance. Therefore, at the loss function level, designing more reasonable loss functions tailored to the characteristics of person re-identification tasks is urgently needed.

Furthermore, current automated machine learning primarily focuses on fundamental tasks such as image classification, object detection, and segmentation. These tasks differ fundamentally from person re-identification tasks, thus extending automated machine learning to the field of person re-identification also presents certain challenges.

1.4 Datasets and Evaluation Metrics

1.4.1 Datasets

According to the settings of this book, the image-based person re-identification datasets will be used for experimental verification in subsequent papers. The datasets used include Market-1501 (Zheng et al. 2015), CUHK03 (Li et al. 2014), and MSMT17 (Wei et al. 2018). The specific relevant statistics of each dataset are shown in Table 1.3.

Market-1501 (Zheng et al. 2015) was collected at Tsinghua University using 6 cameras (5 high-definition cameras and 1 low-definition camera), containing a total of 1501 persons. To address the challenge of cross-camera search, each labeled person appears in at least two cameras. The person bounding boxes were generated using

Table 1.3 Statistics of person re-identification datasets

Dataset	No. of cameras	Training set		Query set		Gallery set	
		Identities	Images	Identities	Images	Identities	Images
Market-1501 (Zheng et al. 2015)	6	751	12936	750	3368	750	15913
CUHK03 (Li et al. 2014)	2	767	7365	700	1400	700	5332
MSMT17 (Wei et al. 2018)	15	1041	32621	3060	11659	3060	82161

the Deformable Part Model (DPM) (Felzenszwalb et al. 2010) to better simulate real-world scenarios. Market-1501 is divided into a training set and a testing set. The training set consists of 12,936 images of 751 persons, with an average of 17.2 images per person. In the testing set, the query set contains 3,368 images of 750 persons with manually drawn bounding boxes. The gallery set of 19,732 images was obtained by DPM detection, which is more consistent with real-world applications. The gallery set includes 3,819 distractor images and 2,798 non-distractor images. As the distractor images do not significantly affect the test results, the final gallery set consists of only the 15,913 non-distractor images.

CUHK03 (Li et al. 2014) was collected at The Chinese University of Hong Kong using 2 cameras, containing a total of 1,467 persons. CUHK03 provides two types of labels: detected and manually labeled. The former introduces detector errors to better fit real-world scenarios, and is therefore more commonly used. The original testing protocol of the CUHK03 dataset differs from that of Market-1501, but the protocol consistent with Market-1501 (Zheng et al. 2015) is now generally adopted. Under the new testing protocol, the training set is composed of 7,365 images of 767 persons, with an average of 9.6 images per person. The testing set consists of a query set with 1,400 images corresponding to 700 persons, and a gallery set with 5,332 images corresponding to 700 persons.

MSMT17 (Wei et al. 2018) was collected at Peking University using 15 cameras, containing a total of 4,101 persons. For this dataset, a person detection model was first used to generate the corresponding detection bounding boxes. The MSMT17 dataset was collected from 12 outdoor cameras and 3 indoor cameras under different weather conditions and at different times of the day, thus making MSMT17 more challenging overall than the previous three datasets. The training set consists of 32,621 images of 1,041 persons, with an average of 31.3 images per person. In the testing set, the query set is composed of 11,659 images corresponding to 3,060 persons, and the gallery set is composed of 82,161 images corresponding to 3,060 persons.

Figure 1.7 shows examples of person images from the Market-1501 and MSMT17 datasets. As can be seen from the figure, images of the same identity vary greatly due to factors such as pose, lighting, viewpoint, occlusion, and detector performance, which poses a great challenge to the task of person re-identification. Although the

Market-1501

MSMT17

Fig. 1.7 Examples of person images from the Market-1501 and MSMT17 datasets

experimental validation in this book is based on these three public datasets, the core key technology is not limited by them. The method proposed in this book can automatically find suitable and effective models for data in real-world application scenarios.

All the experiments in this book were performed in line with the principles of the Declaration of Helsinki. Approval was granted by **Xi'an Research Institute of High Technology**. All person re-identification datasets used in this book are publicly available academic benchmark datasets. The publicly available datasets include Market-1501, CUHK03, and MSMT17.

- **Market-1501**: https://zheng-lab-anu.github.io/Project/project_reid.html
- **CUHK03**: https://www.ee.cuhk.edu.hk/~xgwang/CUHK_identification.html
- **MSMT17**: https://www.pkuvmc.com/dataset.html.

The use of these datasets is limited to non-commercial academic research purposes. The findings presented in this book are intended for scholarly exchange and do not involve any malicious use or infringement upon individual privacy.

1.4.2 Evaluation Metrics

The formalization of person re-identification has been given in Sect. 1.2.2, and a straightforward explanation of performance evaluation has been provided below, combining the Cumulative Matching Characteristics (CMC) curve and the Mean Average Precision (mAP) for a more accurate explanation and definition of model performance.

(1) Cumulative Matching Characteristics Curve. The CMC curve is a commonly used indicator in retrieval tasks, mainly used to calculate the hit rate at Rank-k.

Specifically, through the trained model, the similarity between the query set $\mathcal{Q} = \{(x_i^{(Q)}, y_i^{(Q)})|_{i=1}^{N_Q}\}$ and each image $x_j^{(G)}$ in the gallery set $\mathcal{G} = \{(x_j^{(G)}, y_j^{(G)})|_{j=1}^{N_\mathcal{G}}\}$ is calculated, obtaining the similarity list $[sim(x_i^{(Q)}, x_j^{(\mathcal{G})})|_{j=1}^{N_\mathcal{G}}]$ for each query image $x_i^{(Q)}$. Generally, images in the gallery set $\mathcal{G}$ with the same identity label as the query image $x_i^{(Q)}$ should be removed, as this better reflects the model's ability to track across different scenes. Here, it is assumed that for each query image $x_i^{(Q)}$, all images in the gallery set $\mathcal{G}$ that do not meet the requirements are removed. The similarity list $[sim(x_i^{(Q)}, x_j^{(\mathcal{G})})|_{j=1}^{N_\mathcal{G}}]$ can be expressed as:

$$\text{Rank}k = \frac{1}{N_Q}\sum_{i=1}^{N_Q} \mathbb{I}_{\bigcup_{k'=1}^{k} y_i^{(Q)} = y_{\pi_i(k')}^{(\mathcal{G})}}[1], \tag{1.17}$$

where $\pi_i(k)$ is the value corresponding to the label j of the kth largest entry in the similarity list $[sim(x_i^{(Q)}, x_j^{(\mathcal{G})})|_{j=1}^{N_\mathcal{G}}]$, and $\mathbb{I}_c[o]$ indicates whether the condition c is met, outputting 1 if true and 0 otherwise. The condition is met when the gallery set $\mathcal{G}$ contains the correct sample image $x_i^{(Q)}$ among the k most similar images to the query image $x_i^{(Q)}$. Commonly, Rank-1, Rank-5, and Rank-10 are used to approximate the CMC curve. Figure 1.8 illustrates the calculation process for CMC's Rank-1, Rank-5, and Rank-10, where each blue circle represents the correct sample image of the query image $x_i^{(Q)}$.

(2) Mean Average Precision. mAP is another indicator of ranking performance, which is a supplement to the CMC curve and can more comprehensively reflect the performance of the model. Since the gallery set $\mathcal{G}$ contains not only one image with the same identity label $y_i^{(Q)}$ as the query image $x_i^{(Q)}$, but also other images, the CMC curve only focuses on the first correct image and ignores the subsequent correct images, as shown in Fig. 1.8a, b. The two are consistent in Rank-1, Rank-5, and Rank-10, but the search results in (b) are better. The mAP indicator can better optimize this result, and its specific calculation can be expressed as:

$$\text{mAP} = \frac{1}{N_Q}\sum_{i=1}^{N_Q} AP_i, \tag{1.18}$$

where AP_i represents the average precision for the query image $x_i^{(Q)}$, and can be expressed as:

$$AP_i = \frac{1}{N_{P_i}}\sum_{j=1}^{N_{P_i}} \frac{j}{\eta_{\pi_i}(j)}, \tag{1.19}$$

where $\eta_{\pi_i}(j)$ represents the k corresponding to the jth correct sample image in the sorted list $[sim(x_i^{(Q)}, x_{\pi_i(k)}^{(\mathcal{G})})|_{k=1}^{N_\mathcal{G}}]$ after sorting by π_i. N_{P_i} denotes the number of

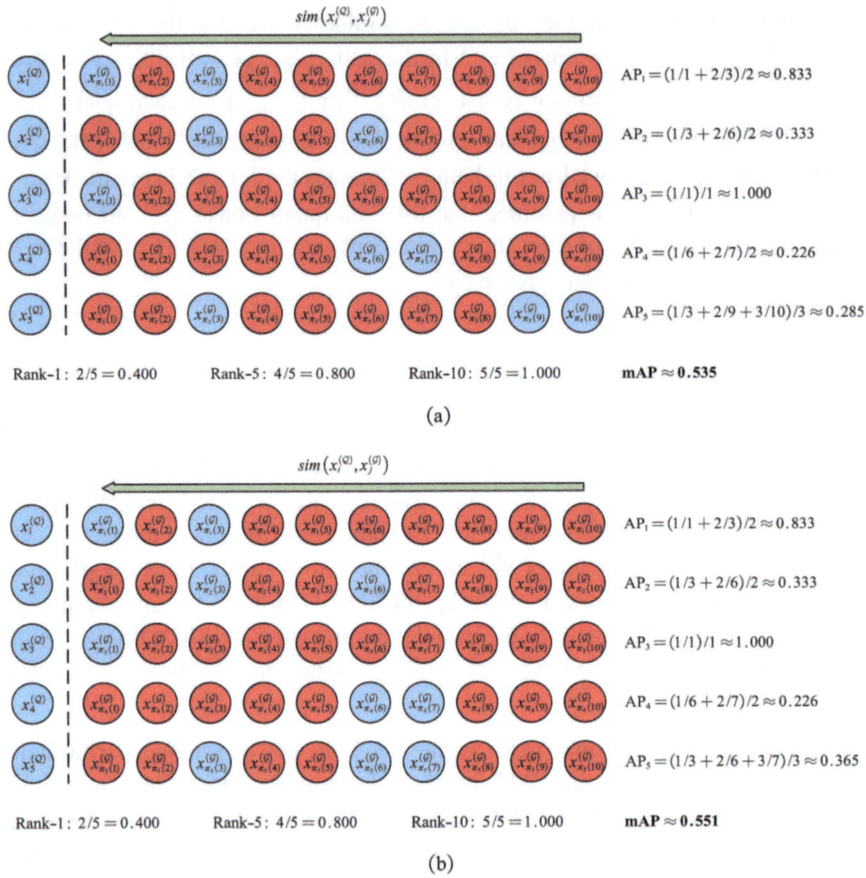

Fig. 1.8 The illustrates examples of the CMC and mAP calculation processes. **a** and **b** are consistent in Rank-1, Rank-5, and Rank-10; however, the mAP of (**b**) is better than that of (**a**), indicating that mAP more comprehensively reflects the ranking performance

correct sample images for the query image $x_i^{(Q)}$ in the gallery set $\mathcal{G}$, which can be expressed as:

$$N_{P_i} = \sum_{j=1}^{N_{\mathcal{G}}} \mathbb{I}_{y_i^{(Q)} = y_j^{(\mathcal{G})}}[1], \tag{1.20}$$

where the definition of $\mathbb{I}_c[o]$ is as per Eq. (1.17). The mAP will only be maximized to 1 when all N_{P_i} correct sample images are ranked in the top N_{P_i} positions in the list $[sim(x_i^{(Q)}, x_{\pi_i(k)}^{(\mathcal{G})})|_{k=1}^{N_{\mathcal{G}}}]$. Therefore, mAP can focus on the positional situation of all correct samples, making it a more comprehensive indicator of model ranking performance than CMC, as shown in Fig. 1.8.

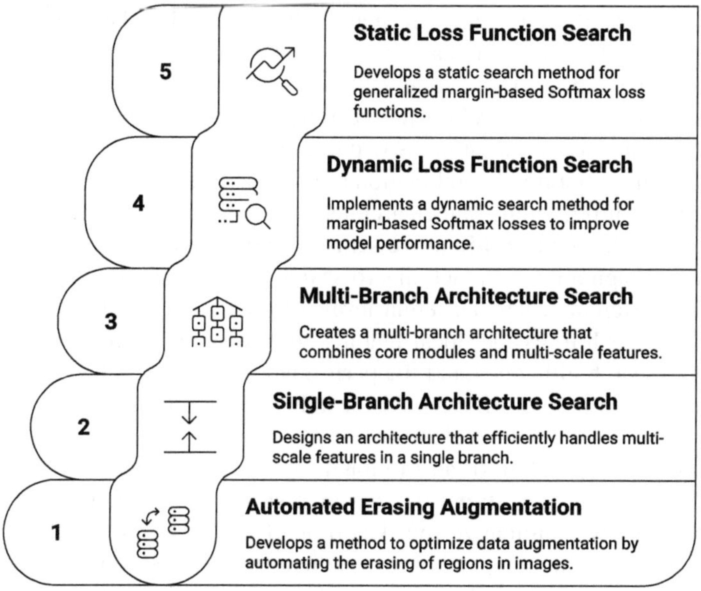

Fig. 1.9 The core research contents of this book

1.5 Research Content and Chapter Arrangement

1.5.1 Research Content

Based on the analysis of the research status and the challenges faced, this book combines automated machine learning (AutoML) techniques to conduct research on person re-identification models based on deep learning, focusing on three aspects: **data augmentation**, **network architecture**, and **loss function**, as shown in Fig. 1.9.

(1) To address the suboptimal generation of erased regions in the current random erasing strategies, we propose a person re-identification method based on **automated erasing data augmentation**. First, we design a differentiable erasing operation that can determine the size, position, and color of the erased region through seven control parameters. Combined with a multivariate Gaussian distribution, a search space is defined. To ensure the differentiability of the distribution sampling process, reparameterization and Cholesky decomposition are employed. A differentiable search algorithm with second-order optimization is utilized to efficiently search the proposed space.
(2) To tackle the suboptimal configuration of multi-scale features in the current single-branch architectures, we propose a person re-identification method based on **automated single-branch architecture search for multi-scale features**. We first design an automatic scaling module composed of lightweight convolutions, enabling different connection configurations of these convolutions to contain

features with different scale information. To ensure that the automatic scaling modules at different levels have distinct and effective structures, we introduce three regularization losses: edge validity, path existence, and graph continuity. Finally, a differentiable architecture search algorithm is used to efficiently search for the connection configurations within each module.

(3) To solve the suboptimal combination of core module structures and multi-scale features in current multi-branch architectures, we propose a **automated multi-scale multi-branch architecture search method** for person re-identification. We first design a three-branch macro-search architecture that includes features of three different scales. Different levels of horizontal partitioning are utilized on each branch to extract local features with varying granularities. Finally, to improve search efficiency and the performance of the searched network architecture, we propose a differentiable search algorithm based on an architecture-stability early-stopping mechanism.

(4) To address the problem of an irrational search space in dynamic loss function search, we propose a **dynamic search method for margin-based Softmax losses**. First, we summarize four key properties of margin-based Softmax loss functions from the perspectives of gradient, features, and optimization. Combined with a non-independent truncated Gaussian distribution, we design a novel loss function sampling distribution. The loss functions sampled from this distribution can satisfy the four key properties mentioned above. Finally, reinforcement learning is used to alternately update the sampling distribution and network weight parameters in a bi-level optimization framework.

(5) To tackle the inefficiency of static loss function search, we propose a **static search method for a generalized margin-based Softmax loss**. First, the generalized margin-based Softmax loss function is represented as two computation graphs and a constant, and a forward construction method is adopted for efficient graph construction. On the basis of regularized evolution, in addition to introducing basic intra-graph and inter-graph mutation operations, we design a loss function search framework that integrates three modules: a loss function rejection criterion, an equivalence verification strategy, and a potential loss function selector. This book aims to provide a comprehensive overview of these novel methods.

1.5.2 Chapter Arrangement

This book comprises a total of seven chapters, each dedicated to addressing the challenges and problems in person re-identification through the design of automated machine learning methods. These methods focus on data augmentation, network architectures, and loss functions. The interrelationships among the chapters are illustrated in Fig. 1.10, and can be summarized as follows:

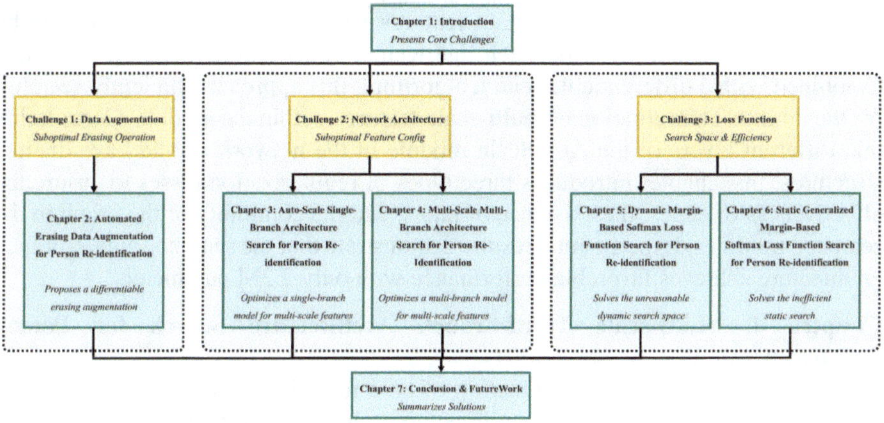

Fig. 1.10 The chapter arrangement of this book

In This Chapter. Introduction
This chapter first introduces the research background and significance of person re-identification. It reviews the current research status in the fields of automatic machine learning and person re-identification based on deep learning. Then, it summarizes the challenges faced by current methods from three aspects: data augmentation, network architecture, and loss function. The relevant datasets and evaluation metrics used in this book are defined and explained. Finally, the five core research contents of this book are outlined in combination with the challenges faced, and the structure of this book is presented.

Chapter 2. Automated Erasing Data Augmentation for Person Re-identification
To address the suboptimal generation of erased regions in the current Random Erasing strategy, this chapter proposes a person re-identification method based on automated erasing data augmentation. This method is built upon a differentiable erasing operation that can determine the size, position, and color of the erased region through seven control parameters, thereby achieving a more flexible way of generating erased regions. Considering the distribution characteristics of the augmentation strategy, this chapter utilizes a multivariate Gaussian distribution and design a transformation that converts the sampling results of the distribution into control parameters for the erasing operation. The search space is defined based on this transformation. By combining the reparameterized sampling process with a second-order optimization-based differentiable search algorithm, this chapter efficiently searches the defined search space.

Chapter 3. Auto-Scale Single-Branch Architecture Search for Person Re-identification
To address the suboptimal configuration of multi-scale features in current single-branch architectures, this chapter proposes a person re-identification method based on automated single-branch architecture search. This method is built upon the designed

AutoScale module, which can obtain features from various scale information by configuring the connections between lightweight convolutions within the module. Combined with a differentiable search algorithm, this approach efficiently searches for the optimal configuration of multi-scale features within the module. Considering that different layers of the AutoScale module in the network should have distinct structures, this chapter introduces three types of regularization losses to ensure the effectiveness of edges, the existence of paths, and the continuity of the graph in the searched modules. Experimental results demonstrate that the searched single-branch architecture achieves favorable performance with only 2.2M parameters.

Chapter 4. Multi-scale Multi-branch Architecture Search for Person Re-identification
To address the suboptimal combination of core module structures and multi-scale features in current multi-branch architectures, this chapter proposes a person re-identification method based on multi-scale multi-branch architecture search. This method is built upon a multi-scale, multi-branch, and global-local macro search architecture. In this architecture, each branch introduces multi-scale features through feature maps of different resolutions and extracts multi-granular local features by combining varying numbers of horizontal divisions. By maintaining the feature scale of each branch, the structures of normal and downsampling modules in the macro search architecture are searched to obtain the optimal structure under the current multi-scale configuration. To improve search efficiency and the performance of the searched network architecture, this chapter introduces a differentiable search algorithm based on an architecture-stability early stopping mechanism.

Chapter 5. Dynamic Margin-based Softmax Loss Function Search for Person Re-identification
To address the issue of the unreasonable search space in current dynamic loss function search methods, this chapter proposes a person re-identification method based on margin-based Softmax dynamic loss search. This method is built upon a search space designed for the margin-based Softmax loss function, where all loss functions satisfy the four key properties analyzed from the perspectives of gradients, features, and optimization. This ensures the rationality of the search space. To dynamically optimize the core part of the search, this chapter employ reinforcement learning to alternately update the sampling distribution and network weights. The sampling distribution is specifically designed for the search space based on a non-independent truncated Gaussian distribution. Once the search is completed, the network weight parameters can be directly used for performance evaluation without the need for retraining.

Chapter 6. Static Generalized Margin-based Softmax Loss Function Search for Person Re-identification
To address the inefficiency in current static loss function search methods, this chapter proposes a person re-identification method based on static loss function search using generalized margin-based Softmax. This method is built upon a search space designed for the generalized margin-based Softmax loss function, where the loss functions are represented as two computational graphs and a constant. The computational

graphs are efficiently constructed using a forward graph construction approach. Based on regularized evolutionary strategies, this chapter designs two types of mutation operations within and between graphs to increase the diversity of offspring. To further enhance search efficiency and the performance of the searched loss functions, this chapter introduces three modules: a loss function rejection criterion, an equivalence verification strategy, and a potential loss function selector. The performance predictor utilized by the potential loss function selector is introduced for the first time in loss function search.

Chapter 7. Conclusion and Future Work
This chapter first summarizes the work presented in this book. It then offers further insights into potential future research directions.

References

Albert Gu and Tri Dao. Mamba: Linear-time sequence modeling with selective state spaces. *CoRR*, abs/2312.00752, 2023.

Alexander Hermans, Lucas Beyer, and Bastian Leibe. In defense of the triplet loss for person re-identification. *CoRR*, abs/1703.07737, 2017.

Alexey Dosovitskiy, Lucas Beyer, Alexander Kolesnikov, Dirk Weissenborn, Xiaohua Zhai, Thomas Unterthiner, Mostafa Dehghani, Matthias Minderer, Georg Heigold, Sylvain Gelly, Jakob Uszkoreit, and Neil Houlsby. An image is worth 16×16 words: Transformers for image recognition at scale. In *9th International Conference on Learning Representations, ICLR 2021, Virtual Event, Austria, May 3–7, 2021*. OpenReview.net, 2021.

Ambika Kaul, Saket Maheshwary, and Vikram Pudi. Autolearn–automated feature generation and selection. In *2017 IEEE International Conference on data mining (ICDM)*, pages 217–226. IEEE, 2017.

Andrea Asperti, Salvatore Fiorilla, Simone Nardi, and Lorenzo Orsini. A review of recent techniques for person re-identification. *Mach. Vis. Appl.*, 36(1):25, 2025.

Barret Zoph and Quoc V. Le. Neural architecture search with reinforcement learning. 2017.

Barret Zoph, Ekin D. Cubuk, Golnaz Ghiasi, Tsung-Yi Lin, Jonathon Shlens, and Quoc V. Le. Learning data augmentation strategies for object detection. In Andrea Vedaldi, Horst Bischof, Thomas Brox, and Jan-Michael Frahm, editors, *Computer Vision - ECCV 2020 - 16th European Conference, Glasgow, UK, August 23-28, 2020, Proceedings, Part XXVII*, volume 12372 of *Lecture Notes in Computer Science*, pages 566–583. Springer, 2020.

Barret Zoph, Vijay Vasudevan, Jonathon Shlens, and Quoc V Le. Learning transferable architectures for scalable image recognition. In *Proceedings of the IEEE conference on computer vision and pattern recognition*, pages 8697–8710, 2018.

Barret Zoph, Vijay Vasudevan, Jonathon Shlens, and Quoc V. Le. Learning transferable architectures for scalable image recognition. In *2018 IEEE Conference on Computer Vision and Pattern Recognition, CVPR 2018, Salt Lake City, UT, USA, June 18–22, 2018*, pages 8697–8710. Computer Vision Foundation / IEEE Computer Society, 2018.

Ben Xie, Xiaofu Wu, Suofei Zhang, Shiliang Zhao, and Ming Li. Learning diverse features with part-level resolution for person re-identification. In Yuxin Peng, Qingshan Liu, Huchuan Lu, Zhenan Sun, Chenglin Liu, Xilin Chen, Hongbin Zha, and Jian Yang, editors, *Pattern Recognition and Computer Vision - Third Chinese Conference, PRCV 2020, Nanjing, China, October 16–18, 2020, Proceedings, Part III*, volume 12307 of *Lecture Notes in Computer Science*, pages 16–28. Springer, 2020.

Bowen Baker, Otkrist Gupta, Nikhil Naik, and Ramesh Raskar. Designing neural network architectures using reinforcement learning. In *5th International Conference on Learning Representations, ICLR 2017, Toulon, France, April 24–26, 2017, Conference Track Proceedings*. OpenReview.net, 2017.

Bryan Bryan, Yuan Gong, Yizhe Zhang, and Christian Poellabauer. Second-order non-local attention networks for person re-identification. In *2019 IEEE/CVF International Conference on Computer Vision, ICCV 2019, Seoul, Korea (South), October 27–November 2, 2019*, pages 3759–3768. IEEE, 2019.

Chanho Eom and Bumsub Ham. Learning disentangled representation for robust person re-identification. In Hanna M. Wallach, Hugo Larochelle, Alina Beygelzimer, Florence d'Alché-Buc, Emily B. Fox, and Roman Garnett, editors, *Advances in Neural Information Processing Systems 32: Annual Conference on Neural Information Processing Systems 2019, NeurIPS 2019, December 8–14, 2019, Vancouver, BC, Canada*, pages 5298–5309, 2019.

Chen Lin, Minghao Guo, Chuming Li, Xin Yuan, Wei Wu, Junjie Yan, Dahua Lin, and Wanli Ouyang. Online hyper-parameter learning for auto-augmentation strategy. In *Proceedings of the IEEE/CVF international conference on computer vision*, pages 6579–6588, 2019.

Guangyi Chen, Tianpei Gu, Jiwen Lu, Jinan Bao, and Jie Zhou. Person re-identification via attention pyramid. *IEEE Trans. Image Process.*, 30:7663–7676, 2021.

Xin Chen, Lingxi Xie, Jun Wu, and Qi Tian. Progressive DARTS: bridging the optimization gap for NAS in the wild. *Int. J. Comput. Vis.*, 129(3):638–655, 2021.

Chenxin Tao, Zizhang Li, Xizhou Zhu, Gao Huang, Yong Liu, and Jifeng Dai. Searching parameterized AP loss for object detection. In Marc'Aurelio Ranzato, Alina Beygelzimer, Yann N. Dauphin, Percy Liang, and Jennifer Wortman Vaughan, editors, *Advances in Neural Information Processing Systems 34: Annual Conference on Neural Information Processing Systems 2021, NeurIPS 2021, December 6–14, 2021, virtual*, pages 22021–22033, 2021.

Chris Thornton, Frank Hutter, Holger H Hoos, and Kevin Leyton-Brown. Auto-weka: Combined selection and hyperparameter optimization of classification algorithms. In *Proceedings of the 19th ACM SIGKDD international conference on Knowledge discovery and data mining*, pages 847–855, 2013.

Christian Szegedy, Vincent Vanhoucke, Sergey Ioffe, Jonathon Shlens, and Zbigniew Wojna. Rethinking the inception architecture for computer vision. In *2016 IEEE Conference on Computer Vision and Pattern Recognition, CVPR 2016, Las Vegas, NV, USA, June 27–30, 2016*, pages 2818–2826. IEEE Computer Society, 2016.

Chuming Li, Xin Yuan, Chen Lin, Minghao Guo, Wei Wu, Junjie Yan, and Wanli Ouyang. Am-lfs: Automl for loss function search. In *Proceedings of the IEEE/CVF international conference on computer vision*, pages 8410–8419, 2019.

Chunfeng Song, Yan Huang, Wanli Ouyang, and Liang Wang. Mask-guided contrastive attention model for person re-identification. In *2018 IEEE Conference on Computer Vision and Pattern Recognition, CVPR 2018, Salt Lake City, UT, USA, June 18–22, 2018*, pages 1179–1188. Computer Vision Foundation / IEEE Computer Society, 2018.

Chunwei Wang, Chao Ma, Ming Zhu, and Xiaokang Yang. Pointaugmenting: Cross-modal augmentation for 3d object detection. In *Proceedings of the IEEE/CVF conference on computer vision and pattern recognition*, pages 11794–11803, 2021.

Connor Shorten and Taghi M Khoshgoftaar. A survey on image data augmentation for deep learning. *Journal of big data*, 6(1):1–48, 2019.

Corinna Cortes and Vladimir Vapnik. Support-vector networks. *Machine learning*, 20:273–297, 1995.

Dangwei Li, Zhang Zhang, Xiaotang Chen, Haibin Ling, and Kaiqi Huang. A richly annotated dataset for pedestrian attribute recognition. *CoRR*, abs/1603.07054, 2016.

Daniel Ho, Eric Liang, Xi Chen, Ion Stoica, and Pieter Abbeel. Population based augmentation: Efficient learning of augmentation policy schedules. In *International conference on machine learning*, pages 2731–2741. PMLR, 2019.

Jiankang Deng, Jia Guo, Jing Yang, Niannan Xue, Irene Kotsia, and Stefanos Zafeiriou. Arcface: Additive angular margin loss for deep face recognition. *IEEE Trans. Pattern Anal. Mach. Intell.*, 44(10):5962–5979, 2022.

Husheng Dong, Ping Lu, Shan Zhong, Chunping Liu, Yi Ji, and Shengrong Gong. Person re-identification by enhanced local maximal occurrence representation and generalized similarity metric learning. *Neurocomputing*, 307:25–37, 2018.

Douglas Gray and Hai Tao. Viewpoint invariant pedestrian recognition with an ensemble of localized features. In David A. Forsyth, Philip H. S. Torr, and Andrew Zisserman, editors, *Computer Vision - ECCV 2008, 10th European Conference on Computer Vision, Marseille, France, October 12–18, 2008, Proceedings, Part I*, volume 5302 of *Lecture Notes in Computer Science*, pages 262–275. Springer, 2008.

Ejaz Ahmed, Michael J. Jones, and Tim K. Marks. An improved deep learning architecture for person re-identification. In *IEEE Conference on Computer Vision and Pattern Recognition, CVPR 2015, Boston, MA, USA, June 7–12, 2015*, pages 3908–3916. IEEE Computer Society, 2015.

Ekin D Cubuk, Barret Zoph, Dandelion Mane, Vijay Vasudevan, and Quoc V Le. Autoaugment: Learning augmentation strategies from data. In *Proceedings of the IEEE/CVF conference on computer vision and pattern recognition*, pages 113–123, 2019.

Ekin D Cubuk, Barret Zoph, Jonathon Shlens, and Quoc V Le. Randaugment: Practical automated data augmentation with a reduced search space. In *Proceedings of the IEEE/CVF conference on computer vision and pattern recognition workshops*, pages 702–703, 2020.

Ekin Dogus Cubuk, Ethan S Dyer, Rapha Gontijo Lopes, and Sylvia Smullin. Tradeoffs in data augmentation: An empirical study. In *Proceedings of the International Conference on Learning Representations*, 2021.

Eric Jang, Shixiang Gu, and Ben Poole. Categorical reparameterization with gumbel-softmax. 2017.

Esteban Real, Chen Liang, David So, and Quoc Le. Automl-zero: Evolving machine learning algorithms from scratch. In *International conference on machine learning*, pages 8007–8019. PMLR, 2020.

Esteban Real, Sherry Moore, Andrew Selle, Saurabh Saxena, Yutaka Leon Suematsu, Jie Tan, Quoc V Le, and Alexey Kurakin. Large-scale evolution of image classifiers. In *International conference on machine learning*, pages 2902–2911. PMLR, 2017.

Esteban Real, Sherry Moore, Andrew Selle, Saurabh Saxena, Yutaka I. Leon-Suematsu, Jie Tan, Quoc V. Le, and Alexey Kurakin. Large-scale evolution of image classifiers. In Doina Precup and Yee Whye Teh, editors, *Proceedings of the 34th International Conference on Machine Learning, ICML 2017, Sydney, NSW, Australia, 6–11 August 2017*, volume 70 of *Proceedings of Machine Learning Research*, pages 2902–2911. PMLR, 2017.

Evelyn Fix. *Discriminatory analysis: nonparametric discrimination, consistency properties*, volume 1. USAF school of Aviation Medicine, 1985.

Fabian Herzog, Xunbo Ji, Torben Teepe, Stefan Hörmann, Johannes Gilg, and Gerhard Rigoll. Lightweight multi-branch network for person re-identification. In *2021 IEEE International Conference on Image Processing, ICIP 2021, Anchorage, AK, USA, September 19–22, 2021*, pages 1129–1133. IEEE, 2021.

Fan Wu, Jinling Gao, Lanqing Hong, Xinbing Wang, Chenghu Zhou, and Nanyang Ye. G-NAS: generalizable neural architecture search for single domain generalization object detection. In Michael J. Wooldridge, Jennifer G. Dy, and Sriraam Natarajan, editors, *Thirty-Eighth AAAI Conference on Artificial Intelligence, AAAI 2024, Thirty-Sixth Conference on Innovative Applications of Artificial Intelligence, IAAI 2024, Fourteenth Symposium on Educational Advances in Artificial Intelligence, EAAI 2014, February 20–27, 2024, Vancouver, Canada*, pages 5958–5966. AAAI Press, 2024.

Xing Fan, Wei Jiang, Hao Luo, and Mengjuan Fei. Spherereid: Deep hypersphere manifold embedding for person re-identification. *J. Vis. Commun. Image Represent.*, 60:51–58, 2019.

Pedro F. Felzenszwalb, Ross B. Girshick, David A. McAllester, and Deva Ramanan. Object detection with discriminatively trained part-based models. *IEEE Trans. Pattern Anal. Mach. Intell.*, 32(9):1627–1645, 2010.

Feng Zheng, Cheng Deng, Xing Sun, Xinyang Jiang, Xiaowei Guo, Zongqiao Yu, Feiyue Huang, and Rongrong Ji. Pyramidal person re-identification via multi-loss dynamic training. In *IEEE Conference on Computer Vision and Pattern Recognition, CVPR 2019, Long Beach, CA, USA, June 16–20, 2019*, pages 8514–8522. Computer Vision Foundation / IEEE, 2019.

Florian Schroff, Dmitry Kalenichenko, and James Philbin. Facenet: A unified embedding for face recognition and clustering. In *IEEE Conference on Computer Vision and Pattern Recognition, CVPR 2015, Boston, MA, USA, June 7-12, 2015*, pages 815–823. IEEE Computer Society, 2015.

Guangyi Chen, Chunze Lin, Liangliang Ren, Jiwen Lu, and Jie Zhou. Self-critical attention learning for person re-identification. In *2019 IEEE/CVF International Conference on Computer Vision, ICCV 2019, Seoul, Korea (South), October 27–November 2, 2019*, pages 9636–9645. IEEE, 2019.

Guanshuo Wang, Yufeng Yuan, Xiong Chen, Jiwei Li, and Xi Zhou. Learning discriminative features with multiple granularities for person re-identification. In Susanne Boll, Kyoung Mu Lee, Jiebo Luo, Wenwu Zhu, Hyeran Byun, Chang Wen Chen, Rainer Lienhart, and Tao Mei, editors, *2018 ACM Multimedia Conference on Multimedia Conference, MM 2018, Seoul, Republic of Korea, October 22-26, 2018*, pages 274–282. ACM, 2018.

Hailin Shi, Yang Yang, Xiangyu Zhu, Shengcai Liao, Zhen Lei, Wei-Shi Zheng, and Stan Z. Li. Embedding deep metric for person re-identification: A study against large variations. In Bastian Leibe, Jiri Matas, Nicu Sebe, and Max Welling, editors, *Computer Vision - ECCV 2016 - 14th European Conference, Amsterdam, The Netherlands, October 11–14, 2016, Proceedings, Part I*, volume 9905 of *Lecture Notes in Computer Science*, pages 732–748. Springer, 2016.

Hanjun Li, Gaojie Wu, and Wei-Shi Zheng. Combined depth space based architecture search for person re-identification. In *IEEE Conference on Computer Vision and Pattern Recognition, CVPR 2021, virtual, June 19–25, 2021*, pages 6729–6738. Computer Vision Foundation / IEEE, 2021.

Hanwen Liang, Shifeng Zhang, Jiacheng Sun, Xingqiu He, Weiran Huang, Kechen Zhuang, and Zhenguo Li. DARTS+: improved differentiable architecture search with early stopping. *CoRR*, abs/1909.06035, 2019.

Hanxiao Liu, Karen Simonyan, and Yiming Yang. Darts: Differentiable architecture search. *arXiv preprint* arXiv:1806.09055, 2018.

Hao Chen, Benoit Lagadec, and François Brémond. Learning discriminative and generalizable representations by spatial-channel partition for person re-identification. In *IEEE Winter Conference on Applications of Computer Vision, WACV 2020, Snowmass Village, CO, USA, March 1–5, 2020*, pages 2472–2481. IEEE, 2020.

Hao Li, Chenxin Tao, Xizhou Zhu, Xiaogang Wang, Gao Huang, and Jifeng Dai. Auto seg-loss: Searching metric surrogates for semantic segmentation. In *9th International Conference on Learning Representations, ICLR 2021, Virtual Event, Austria, May 3–7, 2021*. OpenReview.net, 2021.

Hao Li, Tianwen Fu, Jifeng Dai, Hongsheng Li, Gao Huang, and Xizhou Zhu. Autoloss-zero: Searching loss functions from scratch for generic tasks. In *IEEE/CVF Conference on Computer Vision and Pattern Recognition, CVPR 2022, New Orleans, LA, USA, June 18–24, 2022*, pages 999–1008. IEEE, 2022.

Hao Li, Tianwen Fu, Jifeng Dai, Hongsheng Li, Gao Huang, and Xizhou Zhu. Autoloss-zero: Searching loss functions from scratch for generic tasks. In *Proceedings of the IEEE/CVF Conference on Computer Vision and Pattern Recognition*, pages 1009–1018, 2022.

Hao Wang, Yitong Wang, Zheng Zhou, Xing Ji, Dihong Gong, Jingchao Zhou, Zhifeng Li, and Wei Liu. Cosface: Large margin cosine loss for deep face recognition. In *2018 IEEE Conference on Computer Vision and Pattern Recognition, CVPR 2018, Salt Lake City, UT, USA, June 18–22, 2018*, pages 5265–5274. Computer Vision Foundation / IEEE Computer Society, 2018.

Haowen Xu, Hao Zhang, Zhiting Hu, Xiaodan Liang, Ruslan Salakhutdinov, and Eric P. Xing. Autoloss: Learning discrete schedules for alternate optimization. *CoRR*, abs/1810.02442, 2018.

Haoyuan Sun, Zihao Wu, Bo Xia, Pu Chang, Zibin Dong, Yifu Yuan, Yongzhe Chang, and Xueqian Wang. A method on searching better activation functions. *arXiv preprint* arXiv:2405.12954, 2024.

Xin He, Kaiyong Zhao, and Xiaowen Chu. Automl: A survey of the state-of-the-art. *Knowledge-based systems*, 212:106622, 2021.

Hieu Pham, Melody Y. Guan, Barret Zoph, Quoc V. Le, and Jeff Dean. Efficient neural architecture search via parameter sharing. In Jennifer G. Dy and Andreas Krause, editors, *Proceedings of the 35th International Conference on Machine Learning, ICML 2018, Stockholmsmässan, Stockholm, Sweden, July 10–15, 2018*, volume 80 of *Proceedings of Machine Learning Research*, pages 4092–4101. PMLR, 2018.

Hiroshi Inoue. Data augmentation by pairing samples for images classification. *arXiv preprint* arXiv:1801.02929, 2018.

Houjing Huang, Dangwei Li, Zhang Zhang, Xiaotang Chen, and Kaiqi Huang. Adversarially occluded samples for person re-identification. In *2018 IEEE Conference on Computer Vision and Pattern Recognition, CVPR 2018, Salt Lake City, UT, USA, June 18–22, 2018*, pages 5098–5107. Computer Vision Foundation / IEEE Computer Society, 2018.

Huafeng Qin, Yuming Fu, Jing Chen, Mounim A. El-Yacoubi, Xinbo Gao, and Jun Wang. Neural architecture search based global-local vision mamba for palm-vein recognition. *CoRR*, abs/2408.05743, 2024.

Tao Huang, Zekang Li, Hua Lu, Yong Shan, Shusheng Yang, Yang Feng, Fei Wang, Shan You, and Chang Xu. Relational surrogate loss learning. 2022.

Huantao Zheng, X. Zhong, Wenxin Huang, Kui Jiang, Wenxuan Liu, and Z. Wang. Visible-infrared person re-identification: A comprehensive survey and a new setting. *Electronics*, 2022.

Irwan Bello, Barret Zoph, Vijay Vasudevan, and Quoc V Le. Neural optimizer search with reinforcement learning. In *International Conference on Machine Learning*, pages 459–468. PMLR, 2017.

James Max Kanter and Kalyan Veeramachaneni. Deep feature synthesis: Towards automating data science endeavors. In *2015 IEEE international conference on data science and advanced analytics (DSAA)*, pages 1–10. IEEE, 2015.

Jason Zhi Liang, Santiago Gonzalez, and Risto Miikkulainen. Population-based training for loss function optimization. *CoRR*, abs/2002.04225, 2020.

Jean-Yves Dufour. *Intelligent video surveillance systems*. John Wiley & Sons, 2012.

Jianan Li, Jie Wang, Junjie Chen, and Tingfa Xu. Towards robust point cloud recognition with sample-adaptive auto-augmentation. *IEEE Transactions on Pattern Analysis and Machine Intelligence*, 2025.

Jinxian Liu, Bingbing Ni, Yichao Yan, Peng Zhou, Shuo Cheng, and Jianguo Hu. Pose transferrable person re-identification. In *2018 IEEE Conference on Computer Vision and Pattern Recognition, CVPR 2018, Salt Lake City, UT, USA, June 18–22, 2018*, pages 4099–4108. Computer Vision Foundation / IEEE Computer Society, 2018.

Josif Grabocka, Randolf Scholz, and Lars Schmidt-Thieme. Learning surrogate losses. *CoRR*, abs/1905.10108, 2019.

Juliana Tolles and William J Meurer. Logistic regression: relating patient characteristics to outcomes. *Jama*, 316(5):533–534, 2016.

Kaiming He, Xiangyu Zhang, Shaoqing Ren, and Jian Sun. Deep residual learning for image recognition. In *Proceedings of the IEEE conference on computer vision and pattern recognition*, pages 770–778, 2016.

Keyu Tian, Chen Lin, Ming Sun, Luping Zhou, Junjie Yan, and Wanli Ouyang. Improving autoaugment via augmentation-wise weight sharing. In Hugo Larochelle, Marc'Aurelio Ranzato, Raia Hadsell, Maria-Florina Balcan, and Hsuan-Tien Lin, editors, *Advances in Neural Information Processing Systems 33: Annual Conference on Neural Information Processing Systems 2020, NeurIPS 2020, December 6–12, 2020, virtual*, 2020.

Keyu Tian, Chen Lin, Ser-Nam Lim, Wanli Ouyang, Puneet K. Dokania, and Philip H. S. Torr. A continuous mapping for augmentation design. In Marc'Aurelio Ranzato, Alina Beygelzimer, Yann N. Dauphin, Percy Liang, and Jennifer Wortman Vaughan, editors, *Advances in Neural Information Processing Systems 34: Annual Conference on Neural Information Processing Systems 2021, NeurIPS 2021, December 6–14, 2021, virtual*, pages 13732–13743, 2021.

Jianan Li, Jie Wang, Junjie Chen, and Tingfa Xu. Towards robust point cloud recognition with sample-adaptive auto-augmentation. *IEEE Trans. Pattern Anal. Mach. Intell.*, 47(4):3003–3017, 2025.

Liang Zheng, Liyue Shen, Lu Tian, Shengjin Wang, Jingdong Wang, and Qi Tian. Scalable person re-identification: A benchmark. In *2015 IEEE International Conference on Computer Vision, ICCV 2015, Santiago, Chile, December 7–13, 2015*, pages 1116–1124. IEEE Computer Society, 2015.

Liang Zheng, Yi Yang, and Alexander G. Hauptmann. Person re-identification: Past, present and future. *CoRR*, abs/1610.02984, 2016.

Lijun Wu, Fei Tian, Yingce Xia, Yang Fan, Tao Qin, Lai Jian-Huang, and Tie-Yan Liu. Learning to teach with dynamic loss functions. *Advances in neural information processing systems*, 31, 2018.

Lin Chen, Hua Yang, Shuang Wu, and Zhiyong Gao. Data generation for improving person re-identification. In Qiong Liu, Rainer Lienhart, Haohong Wang, Sheng-Wei "Kuan-Ta" Chen, Susanne Boll, Yi-Ping Phoebe Chen, Gerald Friedland, Jia Li, and Shuicheng Yan, editors, *Proceedings of the 2017 ACM on Multimedia Conference, MM 2017, Mountain View, CA, USA, October 23–27, 2017*, pages 609–617. ACM, 2017.

Lingxi Xie and Alan L. Yuille. Genetic CNN. In *IEEE International Conference on Computer Vision, ICCV 2017, Venice, Italy, October 22–29, 2017*, pages 1388–1397. IEEE Computer Society, 2017.

Lingxiao He, Xingyu Liao, Wu Liu, Xinchen Liu, Peng Cheng, and Tao Mei. Fastreid: A pytorch toolbox for general instance re-identification. In Abdulmotaleb El-Saddik, Tao Mei, Rita Cucchiara, Marco Bertini, Diana Patricia Tobon Vallejo, Pradeep K. Atrey, and M. Shamim Hossain, editors, *Proceedings of the 31st ACM International Conference on Multimedia, MM 2023, Ottawa, ON, Canada, 29 October 2023- 3 November 2023*, pages 9664–9667. ACM, 2023.

Qingliang Liu and Jinmei Lai. Stochastic loss function. In *Proceedings of the AAAI Conference on Artificial Intelligence*, volume 34, pages 4884–4891, 2020.

Liwei Hu, Zidong Wang, Han Li, Peishu Wu, Jingfeng Mao, and Nianyin Zeng. l-darts: Light-weight differentiable architecture search with robustness enhancement strategy. *Knowledge-Based Systems*, 288:111466, 2024.

Longhui Wei, Shiliang Zhang, Wen Gao, and Qi Tian. Person transfer GAN to bridge domain gap for person re-identification. In *2018 IEEE Conference on Computer Vision and Pattern Recognition, CVPR 2018, Salt Lake City, UT, USA, June 18–22, 2018*, pages 79–88. Computer Vision Foundation / IEEE Computer Society, 2018.

Hao Luo, Wei Jiang, Youzhi Gu, Fuxu Liu, Xingyu Liao, Shenqi Lai, and Jianyang Gu. A strong baseline and batch normalization neck for deep person re-identification. *IEEE Trans. Multim.*, 22(10):2597–2609, 2020.

Mahdi M. Kalayeh, Emrah Basaran, Muhittin Gökmen, Mustafa E. Kamasak, and Mubarak Shah. Human semantic parsing for person re-identification. In *2018 IEEE Conference on Computer Vision and Pattern Recognition, CVPR 2018, Salt Lake City, UT, USA, June 18–22, 2018*, pages 1062–1071. Computer Vision Foundation / IEEE Computer Society, 2018.

Mang Ye, Jianbing Shen, Gaojie Lin, Tao Xiang, Ling Shao, and Steven C. H. Hoi. Deep learning for person re-identification: A survey and outlook. *IEEE Trans. Pattern Anal. Mach. Intell.*, 44(6):2872–2893, 2022.

Marc-André Zöller and Marco F Huber. Benchmark and survey of automated machine learning frameworks. *Journal of artificial intelligence research*, 70:409–472, 2021.

Martin Köstinger, Martin Hirzer, Paul Wohlhart, Peter M. Roth, and Horst Bischof. Large scale metric learning from equivalence constraints. In *2012 IEEE Conference on Computer Vision and Pattern Recognition, Providence, RI, USA, June 16–21, 2012*, pages 2288–2295. IEEE Computer Society, 2012.

Matthias Feurer, Katharina Eggensperger, Stefan Falkner, Marius Lindauer, and Frank Hutter. Auto-sklearn 2.0: Hands-free automl via meta-learning. *Journal of Machine Learning Research*, 23(261):1–61, 2022.

Max Jaderberg, Valentin Dalibard, Simon Osindero, Wojciech M Czarnecki, Jeff Donahue, Ali Razavi, Oriol Vinyals, Tim Green, Iain Dunning, Karen Simonyan, et al. Population based training of neural networks. *arXiv preprint* arXiv:1711.09846, 2017.

Maximilian Christ, Nils Braun, Julius Neuffer, and Andreas W Kempa-Liehr. Time series feature extraction on basis of scalable hypothesis tests (tsfresh–a python package). *Neurocomputing*, 307:72–77, 2018.

Zhangqiang Ming, Min Zhu, Xiangkun Wang, Jiamin Zhu, Junlong Cheng, Chengrui Gao, Yong Yang, and Xiaoyong Wei. Deep learning-based person re-identification methods: A survey and outlook of recent works. *Image Vis. Comput.*, 119:104394, 2022.

Mitra Baratchi, Can Wang, Steffen Limmer, Jan N van Rijn, Holger Hoos, Thomas Bäck, and Markus Olhofer. Automated machine learning: past, present and future. *Artificial intelligence review*, 57(5):122, 2024.

Alhassan Mumuni and Fuseini Mumuni. Data augmentation with automated machine learning: approaches and performance comparison with classical data augmentation methods. *Knowl. Inf. Syst.*, 67(5):4035–4085, 2025.

Niki Martinel, Gian Luca Foresti, and Christian Micheloni. Aggregating deep pyramidal representations for person re-identification. In *IEEE Conference on Computer Vision and Pattern Recognition Workshops, CVPR Workshops 2019, Long Beach, CA, USA, June 16–20, 2019*, pages 1544–1554. Computer Vision Foundation / IEEE, 2019.

Niloofar Gheissari, Thomas B. Sebastian, and Richard I. Hartley. Person reidentification using spatiotemporal appearance. In *2006 IEEE Computer Society Conference on Computer Vision and Pattern Recognition (CVPR 2006), 17-22 June 2006, New York, NY, USA*, pages 1528–1535. IEEE Computer Society, 2006.

Peidong Liu, Gengwei Zhang, Bochao Wang, Hang Xu, Xiaodan Liang, Yong Jiang, and Zhenguo Li. Loss function discovery for object detection via convergence-simulation driven search. In *9th International Conference on Learning Representations, ICLR 2021, Virtual Event, Austria, May 3–7, 2021*. OpenReview.net, 2021.

Peng Ye, Baopu Li, Yikang Li, Tao Chen, Jiayuan Fan, and Wanli Ouyang. b-darts: Beta-decay regularization for differentiable architecture search. In *proceedings of the IEEE/CVF conference on computer vision and pattern recognition*, pages 10874–10883, 2022.

Pengguang Chen, Shu Liu, Hengshuang Zhao, Xingquan Wang, and Jiaya Jia. Gridmask data augmentation. *arXiv preprint* arXiv:2001.04086, 2020.

Hieu Pham and Quoc Le. Autodropout: Learning dropout patterns to regularize deep networks. In *Proceedings of the AAAI conference on artificial intelligence*, volume 35, pages 9351–9359, 2021.

Shixi Qin, Zixun Zhang, Yuncheng Jiang, Shuguang Cui, Shenghui Cheng, and Zhen Li. NG-NAS: node growth neural architecture search for 3d medical image segmentation. *Comput. Medical Imaging Graph.*, 108:102268, 2023.

Qiqi Zhou and Yichen Zhu. When training-free nas meets vision transformers: A neural tangent kernel perspective. In *IEEE International Conference on Acoustics, Speech and Signal Processing, ICASSP 2024, Seoul, Republic of Korea, April 14–19, 2024*, pages 7405–7409. IEEE, 2024.

Rahul Rama Varior, Bing Shuai, Jiwen Lu, Dong Xu, and Gang Wang. A siamese long short-term memory architecture for human re-identification. In Bastian Leibe, Jiri Matas, Nicu Sebe, and Max Welling, editors, *Computer Vision - ECCV 2016 - 14th European Conference, Amsterdam, The Netherlands, October 11–14, 2016, Proceedings, Part VII*, volume 9911 of *Lecture Notes in Computer Science*, pages 135–153. Springer, 2016.

Ruibing Hou, Bingpeng Ma, Hong Chang, Xinqian Gu, Shiguang Shan, and Xilin Chen. Interaction-and-aggregation network for person re-identification. In *IEEE Conference on Computer Vision and Pattern Recognition, CVPR 2019, Long Beach, CA, USA, June 16–20, 2019*, pages 9317–9326. Computer Vision Foundation / IEEE, 2019.

Ruihui Li, Xianzhi Li, Pheng-Ann Heng, and Chi-Wing Fu. Pointaugment: An auto-augmentation framework for point cloud classification. In *2020 IEEE/CVF Conference on Computer Vision and Pattern Recognition, CVPR 2020, Seattle, WA, USA, June 13-19, 2020*, pages 6377–6386. Computer Vision Foundation / IEEE, 2020.

Ruijie Quan, Xuanyi Dong, Yu Wu, Linchao Zhu, and Yi Yang. Auto-reid: Searching for a part-aware convnet for person re-identification. In *2019 IEEE/CVF International Conference on Computer Vision, ICCV 2019, Seoul, Korea (South), October 27–November 2, 2019*, pages 3749–3758. IEEE, 2019.

Ryan Layne, Timothy M. Hospedales, and Shaogang Gong. Person re-identification by attributes. In Richard Bowden, John P. Collomosse, and Krystian Mikolajczyk, editors, *British Machine Vision Conference, BMVC 2012, Surrey, UK, September 3–7, 2012*, pages 1–11. BMVA Press, 2012.

Ryuichiro Hataya, Jan Zdenek, Kazuki Yoshizoe, and Hideki Nakayama. Faster autoaugment: Learning augmentation strategies using backpropagation. In *Computer Vision–ECCV 2020: 16th European Conference, Glasgow, UK, August 23–28, 2020, Proceedings, Part XXV 16*, pages 1–16. Springer, 2020.

Sangdoo Yun, Dongyoon Han, Seong Joon Oh, Sanghyuk Chun, Junsuk Choe, and Youngjoon Yoo. Cutmix: Regularization strategy to train strong classifiers with localizable features. In *Proceedings of the IEEE/CVF international conference on computer vision*, pages 6023–6032, 2019.

Santiago Gonzalez and Risto Miikkulainen. Improved training speed, accuracy, and data utilization through loss function optimization. In *IEEE Congress on Evolutionary Computation, CEC 2020, Glasgow, United Kingdom, July 19–24, 2020*, pages 1–8. IEEE, 2020.

Santiago Gonzalez and Risto Miikkulainen. Optimizing loss functions through multi-variate taylor polynomial parameterization. In Francisco Chicano and Krzysztof Krawiec, editors, *GECCO '21: Genetic and Evolutionary Computation Conference, Lille, France, July 10–14, 2021*, pages 305–313. ACM, 2021.

Sasan Salmani Pour Avval, Nathan D Eskue, Roger M Groves, and Vahid Yaghoubi. Systematic review on neural architecture search. *Artificial Intelligence Review*, 58(3):73, 2025.

Robert E. Schapire. The boosting approach to machine learning: An overview. *Springer New York*, 2002.

Shuting He, Hao Luo, Pichao Wang, Fan Wang, Hao Li, and Wei Jiang. Transreid: Transformer-based object re-identification. In *2021 IEEE/CVF International Conference on Computer Vision, ICCV 2021, Montreal, QC, Canada, October 10–17, 2021*, pages 14993–15002. IEEE, 2021.

Sirui Xie, Hehui Zheng, Chunxiao Liu, and Liang Lin. SNAS: stochastic neural architecture search. In *7th International Conference on Learning Representations, ICLR 2019, New Orleans, LA, USA, May 6–9, 2019*. OpenReview.net, 2019.

Stephan Dempe. *Foundations of bilevel programming*. Springer Science & Business Media, 2002.

Sungbin Lim, Ildoo Kim, Taesup Kim, Chiheon Kim, and Sungwoong Kim. Fast autoaugment. *Advances in neural information processing systems*, 32, 2019.

Suofei Zhang, Zirui Yin, Xiaofu Wu, Kun Wang, Quan Zhou, and Bin Kang. FPB: feature pyramid branch for person re-identification. *CoRR*, abs/2108.01901, 2021.

Tianlong Chen, Shaojin Ding, Jingyi Xie, Ye Yuan, Wuyang Chen, Yang Yang, Zhou Ren, and Zhangyang Wang. Abd-net: Attentive but diverse person re-identification. In *2019 IEEE/CVF International Conference on Computer Vision, ICCV 2019, Seoul, Korea (South), October 27–November 2, 2019*, pages 8350–8360. IEEE, 2019.

Pingyu Wang, Zhicheng Zhao, Fei Su, Xingyu Zu, and Nikolaos V. Boulgouris. Horeid: Deep high-order mapping enhances pose alignment for person re-identification. *IEEE Trans. Image Process.*, 30:2908–2922, 2021.

Wei Li, Rui Zhao, Tong Xiao, and Xiaogang Wang. Deepreid: Deep filter pairing neural network for person re-identification. In *2014 IEEE Conference on Computer Vision and Pattern Recognition, CVPR 2014, Columbus, OH, USA, June 23–28, 2014*, pages 152–159. IEEE Computer Society, 2014.

Wei Li, Xiatian Zhu, and Shaogang Gong. Harmonious attention network for person re-identification. In *2018 IEEE Conference on Computer Vision and Pattern Recognition, CVPR*

2018, Salt Lake City, UT, USA, June 18–22, 2018, pages 2285–2294. Computer Vision Foundation / IEEE Computer Society, 2018.

Wei Li, Xiatian Zhu, and Shaogang Gong. Person re-identification by deep joint learning of multi-loss classification. In Carles Sierra, editor, *Proceedings of the Twenty-Sixth International Joint Conference on Artificial Intelligence, IJCAI 2017, Melbourne, Australia, August 19–25, 2017*, pages 2194–2200. ijcai.org, 2017.

Wenjie Yang, Houjing Huang, Zhang Zhang, Xiaotang Chen, Kaiqi Huang, and Shu Zhang. Towards rich feature discovery with class activation maps augmentation for person re-identification. In *IEEE Conference on Computer Vision and Pattern Recognition, CVPR 2019, Long Beach, CA, USA, June 16–20, 2019*, pages 1389–1398. Computer Vision Foundation / IEEE, 2019.

Xiangxiang Chu, Tianbao Zhou, Bo Zhang, and Jixiang Li. Fair DARTS: eliminating unfair advantages in differentiable architecture search. In Andrea Vedaldi, Horst Bischof, Thomas Brox, and Jan-Michael Frahm, editors, *Computer Vision - ECCV 2020 - 16th European Conference, Glasgow, UK, August 23–28, 2020, Proceedings, Part XV*, volume 12360 of *Lecture Notes in Computer Science*, pages 465–480. Springer, 2020.

Xiangxiang Chu, Tianbao Zhou, Bo Zhang, and Jixiang Li. Fair darts: Eliminating unfair advantages in differentiable architecture search. In *European conference on computer vision*, pages 465–480. Springer, 2020.

Xiangyu Zhao, Haochen Liu, Wenqi Fan, Hui Liu, Jiliang Tang, and Chong Wang. Autoloss: Automated loss function search in recommendations. In *Proceedings of the 27th ACM SIGKDD Conference on Knowledge Discovery & Data Mining*, pages 3959–3967, 2021.

Xiaobo Wang, Shuo Wang, Cheng Chi, Shifeng Zhang, and Tao Mei. Loss function search for face recognition. In *Proceedings of the 37th International Conference on Machine Learning, ICML 2020, 13–18 July 2020, Virtual Event*, volume 119 of *Proceedings of Machine Learning Research*, pages 10029–10038. PMLR, 2020.

Xiaolong Wang, Ross B. Girshick, Abhinav Gupta, and Kaiming He. Non-local neural networks. In *2018 IEEE Conference on Computer Vision and Pattern Recognition, CVPR 2018, Salt Lake City, UT, USA, June 18–22, 2018*, pages 7794–7803. Computer Vision Foundation / IEEE Computer Society, 2018.

Xuan Zhang, Hao Luo, Xing Fan, Weilai Xiang, Yixiao Sun, Qiqi Xiao, Wei Jiang, Chi Zhang, and Jian Sun. Alignedreid: Surpassing human-level performance in person re-identification. *CoRR*, abs/1711.08184, 2017.

Xuanyi Dong and Yi Yang. Searching for a robust neural architecture in four GPU hours. In *IEEE Conference on Computer Vision and Pattern Recognition, CVPR 2019, Long Beach, CA, USA, June 16–20, 2019*, pages 1761–1770. Computer Vision Foundation / IEEE, 2019.

Xuelin Qian, Yanwei Fu, Tao Xiang, Wenxuan Wang, Jie Qiu, Yang Wu, Yu-Gang Jiang, and Xiangyang Xue. Pose-normalized image generation for person re-identification. In Vittorio Ferrari, Martial Hebert, Cristian Sminchisescu, and Yair Weiss, editors, *Computer Vision - ECCV 2018 - 15th European Conference, Munich, Germany, September 8–14, 2018, Proceedings, Part IX*, volume 11213 of *Lecture Notes in Computer Science*, pages 661–678. Springer, 2018.

Yang Fu, Yunchao Wei, Yuqian Zhou, Honghui Shi, Gao Huang, Xinchao Wang, Zhiqiang Yao, and Thomas S. Huang. Horizontal pyramid matching for person re-identification. In *The Thirty-Third AAAI Conference on Artificial Intelligence, AAAI 2019, The Thirty-First Innovative Applications of Artificial Intelligence Conference, IAAI 2019, The Ninth AAAI Symposium on Educational Advances in Artificial Intelligence, EAAI 2019, Honolulu, Hawaii, USA, January 27–February 1, 2019*, pages 8295–8302. AAAI Press, 2019.

Yang Yang, Shengcai Liao, Zhen Lei, and Stan Z. Li. Large scale similarity learning using similar pairs for person verification. In Dale Schuurmans and Michael P. Wellman, editors, *Proceedings of the Thirtieth AAAI Conference on Artificial Intelligence, February 12–17, 2016, Phoenix, Arizona, USA*, pages 3655–3661. AAAI Press, 2016.

Kaiwen Yang, Yanchao Sun, Jiahao Su, Fengxiang He, Xinmei Tian, Furong Huang, Tianyi Zhou, and Dacheng Tao. Adversarial auto-augment with label preservation: A representation learning principle guided approach. *Advances in neural information processing systems*, 35:22035–22048, 2022.

Yaoyu Li, Hantao Yao, Tianzhu Zhang, and Changsheng Xu. Part-based structured representation learning for person re-identification. *ACM Trans. Multim. Comput. Commun. Appl.*, 16(4):134:1–134:22, 2021.

Yifan Sun, Changmao Cheng, Yuhan Zhang, Chi Zhang, Liang Zheng, Zhongdao Wang, and Yichen Wei. Circle loss: A unified perspective of pair similarity optimization. In *2020 IEEE/CVF Conference on Computer Vision and Pattern Recognition, CVPR 2020, Seattle, WA, USA, June 13–19, 2020*, pages 6397–6406. Computer Vision Foundation / IEEE, 2020.

Yifan Sun, Liang Zheng, Weijian Deng, and Shengjin Wang. Svdnet for pedestrian retrieval. In *IEEE International Conference on Computer Vision, ICCV 2017, Venice, Italy, October 22–29, 2017*, pages 3820–3828. IEEE Computer Society, 2017.

Yifan Sun, Liang Zheng, Yi Yang, Qi Tian, and Shengjin Wang. Beyond part models: Person retrieval with refined part pooling (and A strong convolutional baseline). In Vittorio Ferrari, Martial Hebert, Cristian Sminchisescu, and Yair Weiss, editors, *Computer Vision - ECCV 2018 - 15th European Conference, Munich, Germany, September 8-14, 2018, Proceedings, Part IV*, volume 11208 of *Lecture Notes in Computer Science*, pages 501–518. Springer, 2018.

Ying Zhang, Baohua Li, Huchuan Lu, Atshushi Irie, and Xiang Ruan. Sample-specific SVM learning for person re-identification. In *2016 IEEE Conference on Computer Vision and Pattern Recognition, CVPR 2016, Las Vegas, NV, USA, June 27–30, 2016*, pages 1278–1287. IEEE Computer Society, 2016.

Yixiao Ge, Zhuowan Li, Haiyu Zhao, Guojun Yin, Shuai Yi, Xiaogang Wang, and Hongsheng Li. FD-GAN: pose-guided feature distilling GAN for robust person re-identification. In Samy Bengio, Hanna M. Wallach, Hugo Larochelle, Kristen Grauman, Nicolò Cesa-Bianchi, and Roman Garnett, editors, *Advances in Neural Information Processing Systems 31: Annual Conference on Neural Information Processing Systems 2018, NeurIPS 2018, December 3–8, 2018, Montréal, Canada*, pages 1230–1241, 2018.

Yonggang Li, Guosheng Hu, Yongtao Wang, Timothy Hospedales, Neil M Robertson, and Yongxin Yang. Differentiable automatic data augmentation. In *Computer Vision–ECCV 2020: 16th European Conference, Glasgow, UK, August 23–28, 2020, Proceedings, Part XXII 16*, pages 580–595. Springer, 2020.

Yuhui Xu, Lingxi Xie, Xiaopeng Zhang, Xin Chen, Guo-Jun Qi, Qi Tian, and Hongkai Xiong. PC-DARTS: partial channel connections for memory-efficient architecture search. In *8th International Conference on Learning Representations, ICLR 2020, Addis Ababa, Ethiopia, April 26–30, 2020*. OpenReview.net, 2020.

Yukang Chen, Tong Yang, Xiangyu Zhang, Gaofeng Meng, Xinyu Xiao, and Jian Sun. Detnas: Backbone search for object detection. In Hanna M. Wallach, Hugo Larochelle, Alina Beygelzimer, Florence d'Alché-Buc, Emily B. Fox, and Roman Garnett, editors, *Advances in Neural Information Processing Systems 32: Annual Conference on Neural Information Processing Systems 2019, NeurIPS 2019, December 8–14, 2019, Vancouver, BC, Canada*, pages 6638–6648, 2019.

Yulin Li, Jianfeng He, Tianzhu Zhang, Xiang Liu, Yongdong Zhang, and Feng Wu. Diverse part discovery: Occluded person re-identification with part-aware transformer. In *IEEE Conference on Computer Vision and Pattern Recognition, CVPR 2021, virtual, June 19–25, 2021*, pages 2898–2907. Computer Vision Foundation / IEEE, 2021.

Yunbiao Zhao, Lei Wang, Jiang Wu, and Taiyong Li. NAST: neural architecture search and transformer for image manipulation localization. *J. Electronic Imaging*, 34(2), 2025.

Zhedong Zheng, Liang Zheng, and Yi Yang. Unlabeled samples generated by GAN improve the person re-identification baseline in vitro. In *IEEE International Conference on Computer Vision, ICCV 2017, Venice, Italy, October 22–29, 2017*, pages 3774–3782. IEEE Computer Society, 2017.

Zhedong Zheng, Xiaodong Yang, Zhiding Yu, Liang Zheng, Yi Yang, and Jan Kautz. Joint discriminative and generative learning for person re-identification. In *IEEE Conference on Computer*

Vision and Pattern Recognition, CVPR 2019, Long Beach, CA, USA, June 16–20, 2019, pages 2138–2147. Computer Vision Foundation / IEEE, 2019.

Zhi-Hua Zhou. *Machine learning*. Springer nature, 2021.

Zhizheng Zhang, Cuiling Lan, Wenjun Zeng, Xin Jin, and Zhibo Chen. Relation-aware global attention for person re-identification. In *2020 IEEE/CVF Conference on Computer Vision and Pattern Recognition, CVPR 2020, Seattle, WA, USA, June 13–19, 2020*, pages 3183–3192. Computer Vision Foundation / IEEE, 2020.

Zhun Zhong, Liang Zheng, Guoliang Kang, Shaozi Li, and Yi Yang. Random erasing data augmentation. In *Proceedings of the AAAI conference on artificial intelligence*, volume 34, pages 13001–13008, 2020.

Zhongwei Zhao, Ran Song, Qian Zhang, Peng Duan, and Youmei Zhang. Jot-gan: A framework for jointly training GAN and person re-identification model. *ACM Trans. Multim. Comput. Commun. Appl.*, 18(1s):27:1–27:18, 2022.

Kaiyang Zhou, Yongxin Yang, Andrea Cavallaro, and Tao Xiang. Learning generalisable omni-scale representations for person re-identification. *IEEE Trans. Pattern Anal. Mach. Intell.*, 44(9):5056–5069, 2022.

Kuan Zhu, Haiyun Guo, Shiliang Zhang, Yaowei Wang, Jing Liu, Jinqiao Wang, and Ming Tang. Aaformer: Auto-aligned transformer for person re-identification. *IEEE Trans. Neural Networks Learn. Syst.*, 35(12):17307–17317, 2024.

Zhun Zhong, Liang Zheng, Donglin Cao, and Shaozi Li. Re-ranking person re-identification with k-reciprocal encoding. In *2017 IEEE Conference on Computer Vision and Pattern Recognition, CVPR 2017, Honolulu, HI, USA, July 21–26, 2017*, pages 3652–3661. IEEE Computer Society, 2017.

Zhun Zhong, Liang Zheng, Zhedong Zheng, Shaozi Li, and Yi Yang. Camera style adaptation for person re-identification. In *2018 IEEE Conference on Computer Vision and Pattern Recognition, CVPR 2018, Salt Lake City, UT, USA, June 18–22, 2018*, pages 5157–5166. Computer Vision Foundation / IEEE Computer Society, 2018.

Zuozhuo Dai, Mingqiang Chen, Xiaodong Gu, Siyu Zhu, and Ping Tan. Batch dropblock network for person re-identification and beyond. In *2019 IEEE/CVF International Conference on Computer Vision, ICCV 2019, Seoul, Korea (South), October 27–November 2, 2019*, pages 3690–3700. IEEE, 2019.

Chapter 2
Automated Erasing Data Augmentation for Person Re-identification

2.1 Introduction

Considering the characteristics of cross-camera retrieval in person re-identification tasks, manual annotation of large datasets is expensive and time-consuming. However, current deep learning-based person re-identification methods still rely on large-scale datasets to prevent network overfitting. An effective way to expand datasets is to employ data augmentation techniques, which are primarily divided into two main categories: generation-based data augmentation and transformation-based data augmentation. The former mainly utilizes generative models to produce new data, but the complexity of their training is often disproportionate to the performance achieved, frequently resulting in diminishing returns. The latter, on the other hand, transforms new data from original data by designing specific rules, and its simplicity and effectiveness have been favored by a wide range of researchers. Currently, transformation-based data augmentation primarily focuses on operations such as random horizontal flipping (RH), random cropping (RC), and random erasing (Zhong et al. 2020) (RE). There is a lack of exploration into other new augmentation operations, which consequently restricts the performance of the entire deep learning-based person re-identification model at the data level.

Recently, with the development of automated machine learning, automated data augmentation based on automated machine learning has achieved many prominent successes (Cubuk et al. 2019; Ho et al. 2019; Wang et al. 2021; Li et al. 2025). Among numerous automated data augmentation methods, differentiable Faster AA (Hataya et al. 2020) and DADA (Li et al. 2020) utilize the Gumbel-Softmax technique (Jang et al. 2017) to approximate non-differentiable operations as differentiable ones, thereby yielding an efficient search algorithm for automated data augmentation strategies. However, most current automated data augmentation methods primarily perform strategy search within the space proposed by AA (Cubuk et al. 2019). A significant portion of this space involves transformations in the image color space, yet color features are critical for person re-identification. Figure 2.1 demonstrates

H. Gu et al., *Automated Machine Learning for Person Re-Identification*,
Intelligent Perception and Information Processing,
https://doi.org/10.1007/978-981-95-3433-3_2

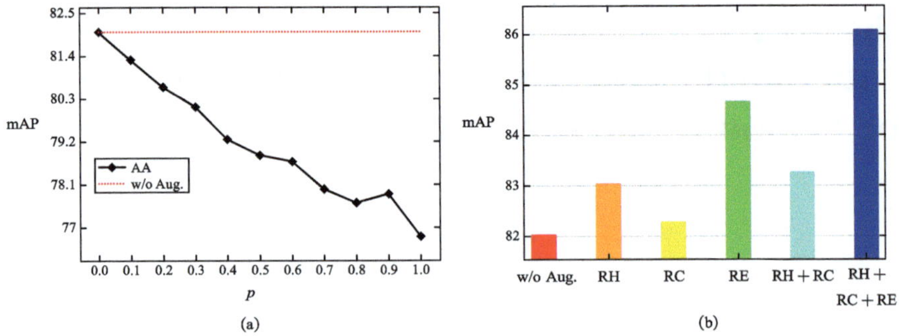

Fig. 2.1 The effectiveness of AA and common data augmentation strategies in the task of person re-identification

the performance of the BoT Luo et al. (2020) model on the Market-1501 dataset using different augmentation strategies. Firstly, as can be seen from the results in Fig. 2.1a, directly applying the augmentation strategies searched by AA on the ImageNet dataset to the person re-identification task yields worse performance than not using any augmentation strategy (w/o Aug.). Furthermore, the performance deteriorates as the probability of this strategy increases, which to some extent verifies that this search space is not friendly to person re-identification tasks. Figure 2.1b illustrates the performance of three common augmentation operations in current person re-identification tasks. It can be observed that these three augmentation operations are effective in improving model performance. When RH+RC+RE are adopted, the model's mAP metric can be improved from 82.02% without augmentation strategies to 86.08%. In addition, Random Erasing among the three operations is the most significant in improving model performance and has become a standard configuration for almost all current deep learning-based person re-identification methods. The Random Erasing strategy primarily has five parameters: controlling the probability of this operation's execution p, the area range of the generated erasing region (s_l, s_h), and the aspect ratio range of this rectangular region (r_l, r_h). From the perspective of the entire erasing process, the Random Erasing strategy has the following issues:

(1) The position of the generated erasing region is completely random. In person re-identification tasks, certain regions (such as the face region (Dietlmeier et al. 2020)) are inherently meaningless, making erasing at all positions suboptimal.
(2) The aspect ratio range of the generated rectangular region is constrained between (r_l, r_h), which limits the mean and diversity of the generated regions. Furthermore, the setting of specific values for (r_l, r_h) is suboptimal.
(3) In Random Erasing, the default three-channel values of 0.485, 0.456, 0.406 are used to fill the rectangular region. The default fill values used across the RGB three channels are suboptimal.
(4) The five control parameters are independent, lacking inherent modeling of their relationships.

To design a data augmentation strategy suitable for person re-identification tasks, and to address the suboptimal generation of erasing regions in the most widely used Random Erasing strategy, this chapter utilizes automated data augmentation technology to propose a person re-identification method based on automated erasing data augmentation—AutoErasing. Firstly, a novel erasing operation is introduced, which can determine the size, position, and color of the rectangular erasing region through seven parameters. Furthermore, the erased image can yield gradients with respect to these seven parameters. Considering the distributional characteristics of augmentation strategies, a transformation is designed that can convert samples from a multivariate Gaussian distribution into values conforming to the ranges of the seven parameters. This defines the search space based on the mean and covariance matrix of this multivariate Gaussian distribution, as well as other parameters. To ensure the differentiability of the multivariate Gaussian sampling process and the positive definiteness of the covariance matrix, the covariance matrix is defined in the form of Cholesky decomposition, and a differentiable multivariate Gaussian sampling process is obtained by combining it with reparameterization techniques. Drawing lessons from current differentiable architecture search algorithms, a second-order optimization search algorithm is employed for efficient searching of this space. This chapter comprehensively experiments and analyzes AutoErasing on three network models (BoT, OSNet, and MGN) across three datasets (Market-1501, CUHK03, and MSMT17). Experimental results demonstrate that the erasing strategies searched by AutoErasing can effectively improve the performance of these three classical network models on the three public datasets.

2.2 Related Works

To meet the demand for large sample sizes in deep models, many large-scale datasets have emerged in person re-identification since 2015, such as the Market-1501 dataset (Zheng et al. 2015) containing 1501 individuals and the MSMT17 dataset (Wei et al. 2018) containing 4101 individuals. Manual annotation of large datasets often requires high costs, thus data augmentation has become a convenient and cost-effective technique for expanding existing datasets. In person re-identification, data augmentation can be mainly divided into two categories: data augmentation based on generation and data augmentation based on transformation.

As a pioneering work in using generative adversarial networks (GANs) for person re-identification data augmentation, LSRO (Zheng et al. 2017) first used deep convolutional generative adversarial networks to generate images. Then, it used a semi-supervised learning framework to provide uniform labels for the generated person images while enhancing the original dataset with the generated person image features. Compared to methods that only use the original dataset, LSRO achieves better performance. Chen et al. (2017) used conditional generative adversarial networks to generate new data samples to alleviate the problem of background interference in the original dataset. Since the background generation only considers the background, the effect is better than LSRO. However, considering the limited improvement in background generation, Pose-transfer (Liu et al. 2018) uses the pose of the original person to generate new data by combining the guidance module, thus improving the quality of the generated images and better considering the background. PN-GAN (Qian et al. 2018) uses a generative adversarial network to eliminate the influence of poses in the original dataset and the generated dataset. It uses two separate deep neural networks to learn the features of the original dataset and the generated dataset, and then uses the original features and the eight typical poses of the generated dataset as the final features. This method can also achieve better performance. In addition to eliminating the influence of poses, camera view differences also need to be considered. CamStyle (Zhong et al. 2018) uses CycleGAN to transfer the images of each camera to the style of other cameras, thus reducing the influence of camera view differences to a certain extent. The above methods treat the generation and feature learning of images as two independent stages, and some methods combine the learning of images and features to promote each other. FD-GAN (Ge et al. 2018) uses the distillation method to integrate the generation and feature learning to obtain identity-related and pose-unrelated features, thus eliminating the influence of pose-related information on feature learning. At the same time, FD-GAN avoids the need for input image pose information in the testing phase, further reducing the dependence on pose detection modules. DG-Net (Zheng et al. 2019) is a task-specific model that learns to generate high-quality person images from human appearance and structural features, thus achieving data augmentation. In addition, DG-Net does not require external information input during both training and testing phases, thus completely avoiding the introduction of noise. However, DG-Net does not fully utilize the feature information of persons, which is not conducive to feature learning. IS-GAN (Eom and Ham 2019) fully learns the features related to and unrelated to persons, thus achieving more prominent performance. JoT-GAN (Zhao et al. 2022) further studies this joint training method and introduces a feature separation module in DG-Net, making feature learning more direct. Although data augmentation based on generation can intuitively obtain generated images, its training complexity and performance do not match, often resulting in a counterproductive effect. Therefore, data augmentation based on transformation, due to its simplicity and convenience, has been favored more.

Data augmentation based on transformation uses predefined rules to transform raw data into new data. In person re-identification tasks, the commonly used paradigms are Random Horizontal Flipping (RH), Random Cropping (RC), and Random Erasing (RE) (Zhong et al. 2020). Among them, Random Erasing erases a rectangular area from the original image with a preset probability p, allowing the model to learn more detailed person features. Specifically, first, a rectangular area size S_e is randomly sampled based on the original image area S and the preset minimum and maximum area values S_l, S_h. Then, the aspect ratio r_e of the rectangular area is randomly sampled based on the preset minimum and maximum aspect ratios r_l, r_h. The length and width of the rectangular area can then be determined as $H_e = \sqrt{S_e \times r_e}$ and $W_e = \sqrt{S_e / r_e}$, respectively. Finally, the center coordinates of the rectangular area (x_e, y_e) are determined based on the length and width H, W of the original image.

The erased rectangular area generated in this way does not guarantee that it will be within the original image. In practice, there will be N attempts to generate the erased area. If none of these N attempts are within the original image, the original image is output. Although Random Erasing is simple, it greatly enhances the robustness of the person re-identification model, so many works consider Random Erasing as a key technique for performance improvement. In addition to the random generation method of Random Erasing, AOS (Huang et al. 2018) uses a pre-trained model to obtain adversarial erased areas and then re-trains the model with these adversarial erased areas. Although AOS can achieve good performance, its training process is more complex and the effect is not prominent. In most cases, using Random Erasing is sufficient. FastReID (He et al. 2023) introduced the augmentation strategy searched by AA into person re-identification. Since this strategy is mainly aimed at ImageNet image classification tasks and contains many operations that are not friendly to person re-identification, such as color space transformations, it actually reduces the model's performance after using this strategy.

Currently, there is still much room for research on data augmentation based on transformation in person re-identification tasks. To avoid hindering person re-identification tasks at the data level, it is necessary to conduct relevant research.

2.3 Methodology

2.3.1 Differentiable Erasing Operation

Figure 2.2 illustrates the proposed differentiable erasing operation. In addition to considering the size and position of the erasing region, its color is also taken into account in the subsequent search. Therefore, seven control parameters (w, h, x, y, r, g, b) are designed to determine the generation of the erasing region. To decouple the erasing operation from the image size, the center of the image is defined as the origin $(0, 0)$, with the top-left corner at $(-1, 1)$ and the bottom-right corner at $(1, 1)$. This allows for erasing in a $[-1, 1]^2$ two-dimensional continuous space, thereby avoiding the influence of the image's dimensions. The width and height of the erasing region are $w, h \in [0, 2]$, respectively, and the coordinates of the center point of this region are $(x, y) \in [-1, 1]^2$. The three-channel color values within the erasing region are $r, g, b \in [0, 1]$.

Given an input image $I \in \mathbb{R}^{3\times H\times W}$, after determining the seven parameters mentioned above, the erasing mask can be expressed as:

$$M[j,i] = \begin{cases} 1, & \text{if } |2i/(W-1) - x| \leqslant w/2 \cap |2j/(H-1) - y| \leqslant h/2 \\ 0, & \text{otherwise} \end{cases}, \quad (2.1)$$

where $M \in \{0, 1\}^{H\times W}$, $i \in \{0, 1, ..., W-1\}$, and $j \in \{0, 1, ..., H-1\}$. Based on this mask, the erased image can be obtained as:

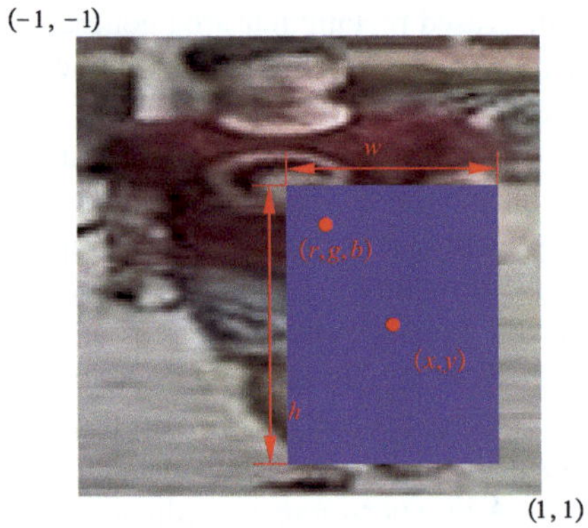

Fig. 2.2 The proposed differentiable erasing operation

$$\begin{aligned} \bar{I} &= I \odot (\mathbf{1} - M) + RGB \odot M, \\ &= I - (I - RGB) \odot M, \end{aligned} \tag{2.2}$$

where $\mathbf{1} \in \{1\}^{H \times W}$ is an all-ones matrix, $RGB \in \mathbb{R}^{3\times1\times1}$ consists of the three values r, g, b in order, and $\odot$ denotes element-wise multiplication. All calculations in the above equations conform to the broadcasting mechanism. It can be observed that the erasing process described above is non-differentiable. The key issue lies in that the mask M in Eq.(2.1) cannot obtain gradients with respect to w, h, x, y. Therefore, we can make the aforementioned process differentiable by designing a soft erasing mask. The soft erasing mask is specifically designed as:

$$\tilde{M}[j, i] = 1 - \max \begin{Bmatrix} \mathrm{Sig}(|2i/(W-1) - 1 - x| - w/2), \\ \mathrm{Sig}(|2j/(H-1) - 1 - y| - h/2) \end{Bmatrix}, \tag{2.3}$$

where $\mathrm{Sig}(x)$ is the Sigmoid function. It can be seen that the soft mask $\tilde{M} \in (0, 1)^{H \times W}$ is related to w, h, x, y and can obtain gradients with respect to them. To ensure that the erased image is consistent with the image erased using Eq.(2.1), the following corrected mask is actually employed:

$$\bar{M} = \mathrm{hs}(\tilde{M}) + \tilde{M} - \mathrm{de}(\tilde{M}), \tag{2.4}$$

where

$$\mathrm{hs}(\tilde{M})[j, i] = \begin{cases} 1, & \text{if } \tilde{M}[j, i] > 0.5 \\ 0, & \text{if } \tilde{M}[j, i] \leqslant 0.5 \end{cases}. \tag{2.5}$$

That is, the soft erasing mask $\tilde{M}$ is converted into a binary mask $\mathrm{hs}(\tilde{M}) \in \{0, 1\}^{H \times W}$ based on a threshold of 0.5, while $\mathrm{de}(x)$ is a function that blocks gradient back-propagation. This design ensures that the corrected mask $\bar{M} \in \{0, 1\}^{H \times W}$ remains consistent with the mask from Eq.(2.1) and can still provide gradients with respect to w, h, x, y.

It should be noted that although the range of w, h is $[0, 2]$, as known from Eq.(2.3), gradients with respect to w, h can also be obtained when w, h are less than 0. The advantage of this is that even if no actual erasing operation is performed on the image, w, h can still be updated based on the corresponding gradients. This avoids the need to design a probability p for operation execution, unlike in random erasing, thereby establishing an intrinsic connection between the various parameters.

2.3.2 Search Space

Unlike neural architecture search and loss function search, automated data augmentation typically requires obtaining a series of augmentation operations, with each operation also involving an execution probability. Based on the differentiable erasing operation described above, the seven control parameters are jointly modeled as follows:

$$\begin{cases} w = 2\tanh(\exp(w') + \theta^{(w)}) \\ h = 2\tanh(\exp(h') + \theta^{(h)}) \\ x = \min\{1, 1 - w/2\} \times \tanh(x') \\ y = \min\{1, 1 - h/2\} \times \tanh(y') \\ r = \mathrm{Sig}(r') \\ g = \mathrm{Sig}(g') \\ b = \mathrm{Sig}(b') \end{cases}, \tag{2.6}$$

where $\mathrm{Sig}(x)$ is the Sigmoid function, and $[w', h', x', y', r', g', b']^T \sim N(\boldsymbol{\mu}, \boldsymbol{\Sigma})$ are the results sampled from a multivariate Gaussian distribution. When both $\theta^{(w)}$ and $\theta^{(h)}$ are equal to 0, the values of w and h are both greater than 0, which means the erasing operation will be performed with 100% probability. Therefore, by introducing $\theta^{(w)}$ and $\theta^{(h)}$, it is possible for w and h to be less than or equal to 0 (in which case no erasing operation is performed). Combined with the designed differentiable erasing operation, even when w, h are less than or equal to 0, gradients with respect to w, h can still be obtained, and these gradients are effective (as demonstrated in subsequent experiments). Thus, the tanh function is used to constrain the range of w, h between $(-2, 2)$, thereby ensuring the inclusion of cases where they are less than or equal to 0. The reason why x, y values are not directly determined by the tanh function is that when the generated erasing region is not contained within the image, it can lead to non-uniqueness, meaning multiple sets of parameters could define the same erased region. To ensure uniqueness, w, h are calculated first, and then the maximum length that can be placed by a region of such size is determined, finally fixing x, y through a certain proportion of $\tanh(x')$ and $\tanh(y')$ under this length. Furthermore, by first determining w, h, cases where no erasing operation occurs are inherently considered, thus eliminating the need to separately consider the implementation probability p. In addition to the parameters $\theta^{(w)}$ and $\theta^{(h)}$, the mean vector $\boldsymbol{\mu}$ and covariance matrix

$\mathbf{\Sigma}$ of the multivariate Gaussian distribution $N(\boldsymbol{\mu}, \mathbf{\Sigma})$ are also parameters that need to be searched for. To ensure the differentiability of the multivariate Gaussian sampling process and the positive definiteness of the covariance matrix, $\mathbf{\Sigma}$ is first expressed using Cholesky decomposition:

$$\mathbf{\Sigma} = LL^T, \tag{2.7}$$

where $L \in \mathbb{R}^{7\times 7}$ is a lower triangular matrix with positive diagonal entries:

$$L = \begin{bmatrix} e^{\theta^{(11)}} & 0 & 0 & 0 & 0 & 0 & 0 \\ \theta^{(21)} & e^{\theta^{(22)}} & 0 & 0 & 0 & 0 & 0 \\ \theta^{(31)} & \theta^{(32)} & e^{\theta^{(33)}} & 0 & 0 & 0 & 0 \\ \theta^{(41)} & \theta^{(42)} & \theta^{(43)} & e^{\theta^{(44)}} & 0 & 0 & 0 \\ \theta^{(51)} & \theta^{(52)} & \theta^{(53)} & \theta^{(54)} & e^{\theta^{(55)}} & 0 & 0 \\ \theta^{(61)} & \theta^{(62)} & \theta^{(63)} & \theta^{(64)} & \theta^{(65)} & e^{\theta^{(66)}} & 0 \\ \theta^{(71)} & \theta^{(72)} & \theta^{(73)} & \theta^{(74)} & \theta^{(75)} & \theta^{(76)} & e^{\theta^{(77)}} \end{bmatrix}. \tag{2.8}$$

This formation ensures the positive definiteness of the covariance. The mean in the Gaussian distribution is expressed as:

$$\boldsymbol{\mu} = \boldsymbol{\theta}^{(\mu)} \in \mathbb{R}^7. \tag{2.9}$$

Then, sampling is performed using the reparameterization trick for multivariate Gaussian distribution:

$$\boldsymbol{x}' = L\boldsymbol{x}_0 + \boldsymbol{\mu}, \tag{2.10}$$

where $\boldsymbol{x}_0 \in \mathbb{R}^7$ represents the result sampled from a standard Gaussian distribution $N(\mathbf{0}, \mathbf{I})$, and $\boldsymbol{x}' = [w', h', x', y', r', g', b']^T$. This sampling method ensures the differentiability of the results sampled from the multivariate Gaussian distribution with respect to the above parameters.

In summary, when considering the search for the color of the erasing region, a total of 37 parameters need to be searched. Without considering color, the total number of parameters is 16. The searches under both these scenarios are the focus of this chapter. Although the proposed erasing strategy in this chapter involves more parameters than the original random erasing, optimizing these parameters allows for more flexible and better-performing erasing strategies.

2.3.3 Search Algorithm

For convenience in subsequent descriptions, the parameters to be searched for are abbreviated as Θ. The current problem can then be formulated as the following bilevel optimization problem:

$$\begin{aligned} \Theta^\star &= \arg\max_{\Theta} \mathcal{P}_V(w^\star(\Theta)), \\ \text{s.t. } w^\star(\Theta) &= \arg\min_{w} \mathcal{L}_T(w, \Theta), \end{aligned} \tag{2.11}$$

where w denotes the weight parameters of the neural network, $w^\star(\Theta)$ represents the optimal weight parameters obtained by the neural network on the training dataset $\mathcal{D}_{trn}$ according to the loss function $\mathcal{L}_T$ under the current augmentation strategy parameters Θ. $\mathcal{P}_V(w^\star(\Theta))$ indicates the performance of the trained neural network on the validation dataset $\mathcal{D}_{val}$. $\Theta^\star$ denotes the optimal erasing strategy configuration to be searched. Since the designed erasing strategy is differentiable with respect to the parameters Θ, a more efficient differentiable search can be considered:

$$\begin{aligned} \Theta^\star &= \arg\min_{\Theta} \mathcal{L}_V(w^\star(\Theta)), \\ \text{s.t. } w^\star(\Theta) &= \arg\min_{w} \mathcal{L}_T(w, \Theta), \end{aligned} \tag{2.12}$$

where $\mathcal{L}_V$ represents the loss function on the validation dataset, which is consistent with $\mathcal{L}_T$. Since there is no independent Θ in $\mathcal{L}_V$, the first-order optimization method commonly used in DARTS (Liu et al. 2018) cannot be directly applied to obtain $\partial\mathcal{L}_V(w^\star(\Theta))/\partial\Theta$. Therefore, a second-order optimization approach is adopted. First, the weights are updated with one iteration:

$$w^{(t)} = w^{(t-1)} - l_w \frac{\partial \mathcal{L}_T(w, \Theta)}{\partial w}\bigg|_{\substack{\Theta=\Theta^{(t-1)} \\ w=w^{(t-1)}}}. \tag{2.13}$$

The weights are updated to approximate the optimal weights $w^\star$, where l_w denotes the weight learning rate and t represents the iteration number. For convenience in subsequent descriptions, let $\nabla_\Theta \mathcal{L}(\Theta) \triangleq \partial\mathcal{L}(\Theta)/\partial\Theta$. Then, we have:

$$\begin{aligned} \nabla_\Theta \mathcal{L}_T(w^\star(\Theta))\Big|_{\Theta=\Theta^{(t-1)}} &\approx \nabla_\Theta \mathcal{L}_V(w - l_w \nabla_w \mathcal{L}_T(w, \Theta))\Big|_{\substack{\Theta=\Theta^{(t-1)} \\ w=w^{(t-1)}}} \\ &= -l_w \nabla^2_{\Theta,w} \mathcal{L}_T(w, \Theta)\Big|_{\substack{\Theta=\Theta^{(t-1)} \\ w=w^{(t-1)}}} \nabla_w \mathcal{L}_V(w)\Big|_{w=w^{(t)}}. \end{aligned} \tag{2.14}$$

Since solving for $\nabla^2_{\Theta,w} \mathcal{L}_T(w, \Theta)$ consumes excessive GPU resources, a finite difference approximation is adopted for its approximate computation:

$$\begin{aligned} \nabla_\Theta \mathcal{L}_T(w^\star(\Theta))\Big|_{\Theta=\Theta^{(t-1)}} &\approx -l_w \nabla^2_{\Theta,w} \mathcal{L}_T(w, \Theta)\Big|_{\substack{\Theta=\Theta^{(t-1)} \\ w=w^{(t-1)}}} \nabla_w \mathcal{L}_V(w)\Big|_{w=w^{(t)}} \\ &\approx -l_w \Big(\nabla_\Theta \mathcal{L}_T(w^{(t-1)} + \epsilon \nabla_w \mathcal{L}_V(w)\Big|_{w=w^{(t)}}, \Theta)\Big|_{\Theta=\Theta^{(t-1)}} \\ &\quad - \nabla_\Theta \mathcal{L}_T(w^{(t-1)} - \epsilon \nabla_w \mathcal{L}_V(w)\Big|_{w=w^{(t)}}, \Theta)\Big|_{\Theta=\Theta^{(t-1)}} \Big) / (2\epsilon), \end{aligned} \tag{2.15}$$

where ϵ is a small value. This leads to the update rule for Θ:

$$\Theta^{(t)} = \Theta^{(t-1)} - l_{\Theta} \frac{\partial \mathcal{L}_V(w^{\star}(\Theta))}{\partial \Theta}\bigg|_{\Theta=\Theta^{(t-1)}}, \tag{2.16}$$

where l_{Θ} denotes the strategy parameter learning rate. The proposed AutoErasing automated data augmentation algorithm is presented in Algorithm 1.

Algorithm 1 AutoErasing Automated Data Augmentation Algorithm

Require: Erasing parameters Θ; Weight parameters w; Training dataset $\mathcal{D}_T$; Total update epochs $Epoch$; Number of updates per epoch I
Ensure: Optimal configuration Θ^*
Divide the training dataset $\mathcal{D}_T$ into search training set $\mathcal{D}_{trn}$ and search validation set $\mathcal{D}_{val}$;
for $e = 1$ to $Epoch$ **do**
  for $i = 1$ to I **do**
    Fetch a batch of data from $\mathcal{D}_{trn}$;
    Update w according to Eq.(2.13);
    Fetch a batch of data from $\mathcal{D}_{val}$;
    Update Θ according to Eq.(2.16);
  end for
end for
Return $\Theta^* = \Theta$.

2.4 Experiments

2.4.1 Datasets and Evaluation Metrics

The performance of the searched erasing strategies is evaluated using CMC curves and mAP metrics on three commonly used datasets (Market-1501 (Zheng et al. 2015), CUHK03 (Li et al. 2014), and MSMT17 (Wei et al. 2018)) and three classic network architectures (BoT (Luo et al. 2020), OSNet (Zhou et al. 2022), and MGN (Wang et al. 2018)). Firstly, the basic experimental setup is presented. Then, ablation experiments are conducted on AutoErasing. Finally, a comparison of the three classic models with other state-of-the-art methods is provided, integrating the searched erasing strategies for this chapter.

2.4.2 Implementation Details

During the search phase, the original training data $\mathcal{D}_T$ in the dataset is divided into search training dataset $\mathcal{D}_{trn}$ and search validation dataset $\mathcal{D}_{val}$ with a 1:1 ratio. The erasing strategies are searched on the Market-1501 dataset in conjunction with the BoT model. Subsequently, the discovered strategies are directly transferred to the other three datasets, as well as the OSNet and MGN models. Two distinct search spaces are primarily explored: a search without considering color channels, where the total number of parameters is 16, and the strategies found in this case are named AE; and a search considering color channels, with a total of 37 parameters, where the strategies found are named AE-C. The input image size is set to 256×128, and batch data is constructed using PK sampling, with $P = 16$ and $K = 4$. The total number of update epochs is 200, and the number of updates per epoch, I, depends on PK sampling and the size of the $\mathcal{D}_{trn}$ dataset. An SGD optimizer with momentum (0.9) is employed to update the network weight parameters w. The weight learning rate l_w is initialized to 0.05 and gradually reduced to 0.0005 using a cosine annealing schedule (Loshchilov and Hutter 2017). The erasing parameters Θ are updated using an Adam optimizer with a learning rate of $l_\Theta = 0.0005$. The weight decay for SGD is set to 0.0005. L_r and L_c are consistent with BoT (Luo et al. 2020). The search time requires approximately 1.0 day on a single NVIDIA 2080Ti GPU, and all code for this chapter is based on PyTorch (Paszke et al. 2019).

During the re-training phase, for BoT and OSNet, the input image size is set to 256×128, while for MGN's input image size is set to 384×128. Data augmentation strategies commonly employed include random cropping, random horizontal flipping, and the searched erasing strategies. Batch data is constructed using $P - K$ sampling, with P and K set to 16 and 4, respectively. The specific training configurations for the three models, BoT (Luo et al. 2020), OSNet (Zhou et al. 2022), and MGN (Wang et al. 2018), remain consistent with their original models. During testing, test-time horizontal flipping is applied, and cosine similarity is used as the metric for similarity measurement.

2.4.3 Comparison with State-of-the-Art Methods

Table 2.1 presents the comparison results of the searched erasing strategy combined with classic BoT, OSNet, and MGN, as well as other state-of-the-art methods. The double-line section shows the results based on global feature models, while the lower part presents the results based on global-local/multi-branch models. Within each section, below the single line, the performance of the erasing strategy searched by AutoErasing on the corresponding model is shown. R1 denotes the Rank-1 metric. **Bold metrics** indicate the best performance of the searched strategy on the two types of models, while underlined metrics represent the best performance of other methods on the two types of models. The results show that the searched erasing strategy can effectively improve the performance of these three classic models, compared to the original ResNet-50-based approach.

For BoT, even the AE strategy without considering color space can improve mAP and Rank-1 by 0.7% and 0.25%, respectively, on three datasets. On the OSNet model, mAP and Rank-1 metrics were improved by an average of 0.83% and 0.15%, respectively. And on the MGN model, mAP and Rank-1 metrics were improved by an average of 0.5% and 0.2%, respectively.

Table 2.1 Comparison of AutoErasing with state-of-the-art methods on three datasets

Method	Market-1501		CUHK03		MSMT17	
	mAP	R1	mAP	R1	mAP	R1
TriNet (Hermans et al. 2017)	69.1	84.9	–	–	–	–
SphereReID (Fan et al. 2019)	83.6	94.4	–	–	–	–
BoT (Luo et al. 2020)	85.9	94.5	–	–	–	–
ResNet-50+CircleLoss (Sun et al. 2020)	84.9	94.2	–	–	50.2	76.3
OSNet (Zhou et al. 2022)	86.7	94.8	67.8	72.3	55.1	79.1
Auto-ReID (Quan et al. 2019)	85.1	94.5	69.3	73.3	52.5	78.2
BoT+AE	86.7	94.4	59.9	61.1	52.0	75.5
BoT+AE-C	87.0	94.6	60.0	62.3	53.0	76.6
OSNet+AE	87.2	95.0	68.4	72.7	55.7	79.5
OSNet+AE-C	**87.9**	**95.3**	**69.5**	**73.5**	**56.4**	**79.7**
MGN (Wang et al. 2018)	86.9	95.7	66.0	66.8	–	–
MGN+CircleLoss (Sun et al. 2020)	87.4	96.1	–	–	52.1	76.9
ABD-Net (Chen et al. 2019)	88.3	95.6	–	–	60.8	82.3
MGN+AE	87.4	95.4	66.8	67.5	59.0	80.8
MGN+AE-C	**88.0**	**95.5**	**67.3**	**67.9**	**60.3**	**82.3**

When the AE-C strategy considering color space is adopted, compared to the original ResNet-50-based BoT, the AE-C strategy can improve mAP and Rank-1 by 1.3% and 0.85%, respectively, on three datasets. On the OSNet model, mAP and Rank-1 metrics were improved by an average of 1.58% and 0.63%, respectively. And on the MGN model, mAP and Rank-1 metrics were improved by an average of 1.07% and 0.47%, respectively.

Comparing the two searched strategies, AE and AE-C, the AE-C strategy, which considers color space, yields models with better generalization performance than the AE strategy that does not consider color space. Furthermore, the method proposed in this chapter primarily focuses on expanding person re-identification datasets at the data level, which is orthogonal to work like ABD-Net (Chen et al. 2019) that deals with network architecture design. Therefore, applying the proposed method to these models can achieve even better performance.

2.4.4 Ablation Study

(1) The Effectiveness of Differentiable Erasing Operation. Firstly, to verify the correctness of the designed differentiable erasing operation, we directly optimized the seven parameters w, h, x, y, r, g, b. The image after random erasing was used as the erased image. The Mean Square Error (MSE) loss was employed:

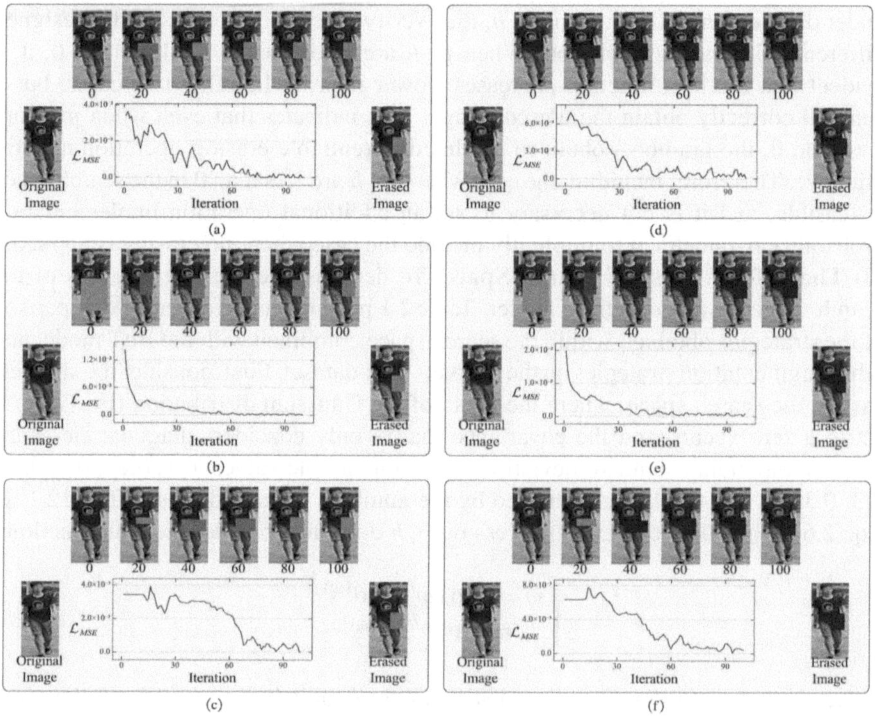

Fig. 2.3 The effectiveness of differentiable erasing operation on different initializations of w, h

$$\mathcal{L}_{\mathrm{MSE}} = \frac{1}{3HW}\sum_{i,j,c}\left(\overline{I}[c,i,j] - \mathrm{DE}(I;\alpha)[c,i,j]\right)^2, \tag{2.17}$$

where $I \in \mathbb{R}^{3\times H\times W}$ is the original image without erasing, as shown as "Original Image" in Fig. 2.3. $\overline{I} \in \mathbb{R}^{3\times H\times W}$ is the erased image obtained by applying random erasing to the original image I using Eq. (2.2), as shown as "Erased Image" in Fig. 2.3. $DE(I;\alpha)$ is the designed differentiable erasing operation, and α is its parameter. When using the default erasing region color of 0.485, 0.456, 0.406, $\alpha = \{w, h, x, y\}$, as shown in Fig. 2.3a–c; when color channels are considered, $\alpha = \{w, h, x, y, r, g, b\}$, as shown in Fig. 2.3d–f. To further illustrate that the designed operation can still yield effective gradients when w, h are negative, three different initializations for w, h were designed: namely, w, h initialized to 0 (as shown in Fig. 2.3a, d), w, h initialized to be greater than 0 (as shown in Fig. 2.3b, e), and w, h initialized to be less than 0 (as shown in Fig. 2.3c, f). All 6 experiments above were conducted using the same random seed, employing the Adam optimizer with a learning rate of 0.1 for a total of 100 iterations.

From the experimental results in Fig. 2.3, it can be seen that the differentiable erasing operation designed in this chapter can correctly obtain the erased image under different initializations of w, h, thus verifying the correctness of the designed differentiable erasing operation. When w, h are initialized to be less than 0, it is evident that the loss function decreases slower than in the other two cases, but it can still correctly obtain the erased image. This indicates that even when w, h are less than 0, the gradients obtained by this differentiable erasing operation are still effective. Therefore, including the case where w, h are less than 0 in the search space is feasible, and it is not necessary to set an additional operation implementation probability parameter p to implicitly include the case where no erasing is applied.

(2) The Effectiveness of Search Space. To demonstrate the effectiveness of the search space designed in this chapter, Table 2.1 presents a performance comparison of the strategies obtained within the search space combined with the BoT model and other augmentation strategies on the Market-1501 dataset. First, consider the simplest case in the search space, where the mean of the Gaussian distribution (Eq. (2.9)) is set to a zero vector, and the covariance matrix only considers diagonal elements. The corresponding standard deviations are set to simple values: 0.1, 0.15, 0.2, 0.25, 0.3, 0.35, 0.4, 0.45, 0.5, as indicated by the numbers in parentheses in Table 2.1. In Eq. (2.6), when the generation process of w, h does not consider the tanh function:

$$\begin{cases} \tilde{w} = \exp(w') + \theta^{(w)} \\ \tilde{h} = \exp(h') + \theta^{(h)} \end{cases} . \tag{2.18}$$

Since w' and h' follow Gaussian distributions, $\tilde{w}$ and $\tilde{h}$, respectively, follow log-normal distributions with offsets $\theta^{(w)}$ and $\theta^{(h)}$. Generally, when designing augmentation strategies, it is desirable that the resulting strategy has a high probability of not applying any operation. Therefore, $\theta^{(w)}$ and $\theta^{(h)}$ are set to $-\exp(-\sigma^2)$ (where σ is the set standard deviation) to meet the above requirements. The effectiveness of the search space was analyzed primarily by combining two scenarios: one where AE does not consider color space, and another where AE-C considers color space. In the table, both AE and AE-C utilize Random Horizontal Flipping (RH) + Random Cropping (RC), which are not explicitly indicated in Table 2.2 to save space. W/o Aug. denotes the absence of any augmentation strategy.

From the results in Table 2.2, it can be seen that AE(0.3) in the search space can achieve performance close to that of Random Erasing (RE) fine-tuned to its optimum (as indicated by RE* in Table 2.2). This demonstrates the effectiveness of the proposed search space. Furthermore, AE-C(0.3), which considers color space, can achieve an mAP of 86.52%, indicating that strategies considering color space can achieve better performance. Finally, the optimal results found in these two search spaces are presented (as indicated by AE* and AE-C* in Table 2.2). AE* achieves an mAP of 86.72%, while AE-C* reaches an mAP of 87.00%. This further validates the effectiveness of the proposed search space.

(3) The Visualization of the Searched Erasing Strategies. Figure 2.4 shows the visualization of the searched strategies when color space is not considered. By randomly sampling 10,000 times from the distribution of the searched strategies, we obtain the corresponding w, h, x, y values. The distributions of x and y only con-

Table 2.2 Comparison of AutoErasing with other common augmentation strategies

Augmentation strategy	mAP	Rank-1	Augmentation strategy	mAP	Rank-1	Augmentation strategy	mAP	Rank-1
W/o Aug.	82.02	93.20	AE(0.10)	83.46	94.18	AE-C(0.10)	84.17	94.30
RH	83.03	93.91	AE(0.15)	84.40	94.15	AE-C(0.15)	84.73	94.30
RC	82.28	93.68	AE(0.20)	85.34	94.74	AE-C(0.20)	85.64	95.13
RE	84.66	94.00	AE(0.25)	86.03	95.19	AE-C(0.25)	86.27	94.92
RH+RC	83.26	94.24	AE(0.30)	86.36	94.92	AE-C(0.30)	86.52	94.98
RH+RE	85.94	94.33	AE(0.35)	86.12	94.27	AE-C(0.35)	86.23	94.34
RC+RE	84.90	94.19	AE(0.40)	85.83	94.18	AE-C(0.40)	86.00	94.28
RH+RC+RE	86.08	94.39	AE(0.45)	85.45	94.15	AE-C(0.45)	85.72	94.38
–	–	–	AE(0.50)	84.96	94.06	AE-C(0.50)	85.19	94.17
RH+RC+RE*	86.44	94.41	AE*	86.72	94.57	AE-C*	87.00	94.63

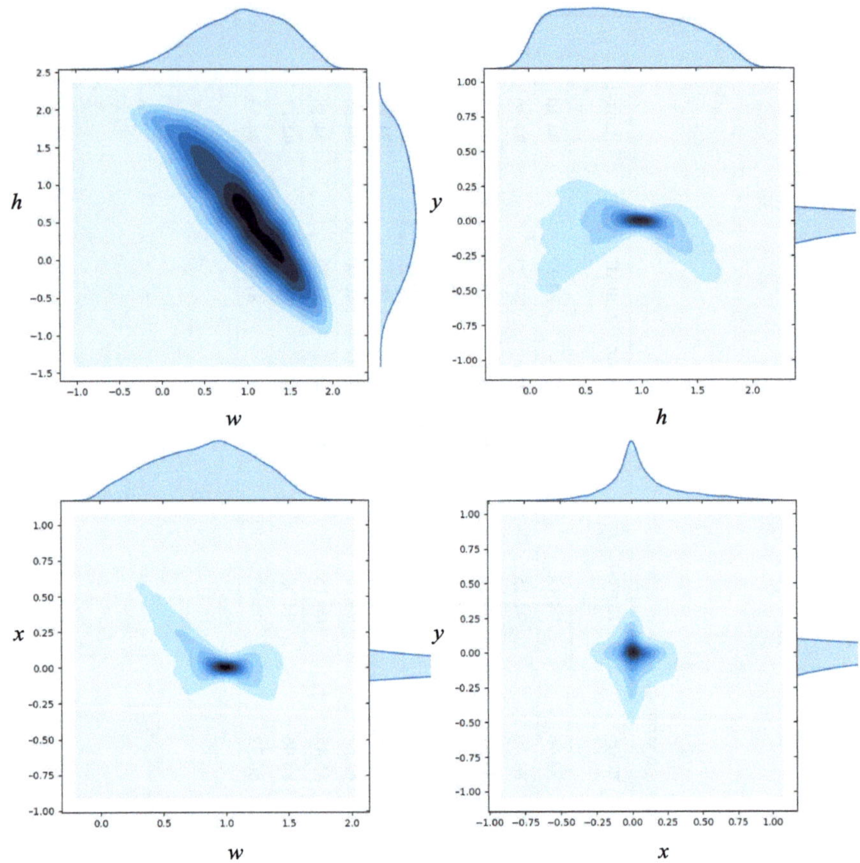

Fig. 2.4 The visualization of the searched AE strategies

sider cases where both w and h are greater than 0, as an effective erasing region only exists and x, y values are meaningful under this condition. By summing the number of sample points where w and h are greater than 0 in the top-left distribution and dividing by 10,000, an approximate probability of 0.7562 for the erasing operation to be applied can be obtained. This is slightly higher than the current default probability of 0.5, indicating that the current default probability is relatively small. The positions of the original random erasing are uniformly distributed across all locations. From the visualization results of the distribution, it can be seen that there is a certain correlation between the position and the size of the erasing region. Furthermore, the x, y distribution in the bottom-right indicates that the generated erasing regions are more inclined towards $y < 0$, i.e., the upper part of the image. Intuitively, the upper body clothing of a person has many important features, so overly strong occlusion in this part can easily lead to model overfitting. From the distribution in the top-right, when

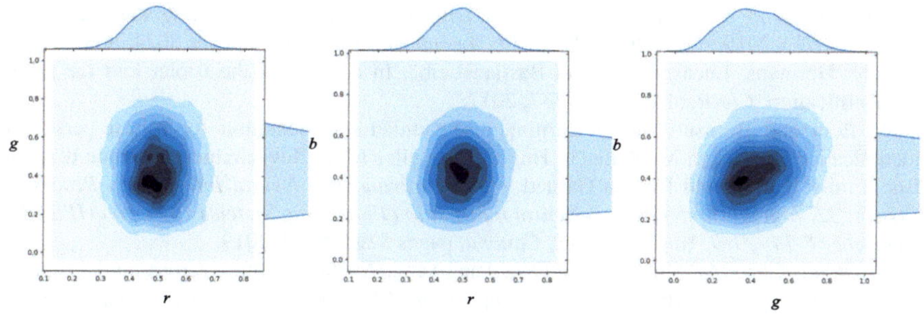

Fig. 2.5 The visualization of the searched AE-C strategies

y is in the region of -0.5, h is smaller. Therefore, it is reasonable that the searched strategy adopts a smaller erasing region for these important feature areas.

Figure 2.5 shows the visualization of the three color channels for the AE-C strategy. The distributions of the other four parameters are similar to those of AE, and thus will not be elaborated upon. An analysis of the results in Fig. 2.5 reveals a significant correlation between g and b in the three color channels. Moreover, while the r value in the three channels is centered around 0.5, the other two values tend to be centered around 0.4. This indicates that the three channels have learned certain information, and from the results in Table 2.1, it can also be seen that strategies that consider color space perform better than those that do not.

2.5 Conclusion

This chapter, from the perspective of automated data augmentation, proposes a person re-identification method based on automated erasing data augmentation—AutoErasing. This method addresses the suboptimal generation of erasing regions in currently common random erasing strategies, by conducting an automated data augmentation strategy search for the size, position, and color of erasing regions. The searched strategies can effectively improve the performance of multiple classic models on three common datasets, providing a feasible approach for designing effective data augmentation in person re-identification.

References

Adam Paszke, Sam Gross, Francisco Massa, Adam Lerer, James Bradbury, Gregory Chanan, Trevor Killeen, Zeming Lin, Natalia Gimelshein, Luca Antiga, Alban Desmaison, Andreas Köpf, Edward Z. Yang, Zachary DeVito, Martin Raison, Alykhan Tejani, Sasank Chilamkurthy, Benoit Steiner, Lu Fang, Junjie Bai, and Soumith Chintala. Pytorch: An imperative style, high-performance deep learning library. In Hanna M. Wallach, Hugo Larochelle, Alina Beygelzimer, Florence d'Alché-Buc, Emily B. Fox, and Roman Garnett, editors, *Advances in Neural Infor-*

mation Processing Systems 32: Annual Conference on Neural Information Processing Systems 2019, NeurIPS 2019, December 8-14, 2019, Vancouver, BC, Canada, pages 8024–8035, 2019.

Alexander Hermans, Lucas Beyer, and Bastian Leibe. In defense of the triplet loss for person re-identification. *CoRR*, abs/1703.07737, 2017.

Chanho Eom and Bumsub Ham. Learning disentangled representation for robust person re-identification. In Hanna M. Wallach, Hugo Larochelle, Alina Beygelzimer, Florence d'Alché-Buc, Emily B. Fox, and Roman Garnett, editors, *Advances in Neural Information Processing Systems 32: Annual Conference on Neural Information Processing Systems 2019, NeurIPS 2019, December 8-14, 2019, Vancouver, BC, Canada*, pages 5298–5309, 2019.

Chunwei Wang, Chao Ma, Ming Zhu, and Xiaokang Yang. Pointaugmenting: Cross-modal augmentation for 3d object detection. In *Proceedings of the IEEE/CVF conference on computer vision and pattern recognition*, pages 11794–11803, 2021.

Daniel Ho, Eric Liang, Xi Chen, Ion Stoica, and Pieter Abbeel. Population based augmentation: Efficient learning of augmentation policy schedules. In *International conference on machine learning*, pages 2731–2741. PMLR, 2019.

Ekin D Cubuk, Barret Zoph, Dandelion Mane, Vijay Vasudevan, and Quoc V Le. Autoaugment: Learning augmentation strategies from data. In *Proceedings of the IEEE/CVF conference on computer vision and pattern recognition*, pages 113–123, 2019.

Eric Jang, Shixiang Gu, and Ben Poole. Categorical reparameterization with gumbel-softmax. 2017.

Guanshuo Wang, Yufeng Yuan, Xiong Chen, Jiwei Li, and Xi Zhou. Learning discriminative features with multiple granularities for person re-identification. In Susanne Boll, Kyoung Mu Lee, Jiebo Luo, Wenwu Zhu, Hyeran Byun, Chang Wen Chen, Rainer Lienhart, and Tao Mei, editors, *2018 ACM Multimedia Conference on Multimedia Conference, MM 2018, Seoul, Republic of Korea, October 22-26, 2018*, pages 274–282. ACM, 2018.

Hanxiao Liu, Karen Simonyan, and Yiming Yang. Darts: Differentiable architecture search. *arXiv preprint* arXiv:1806.09055, 2018.

Hao Luo, Wei Jiang, Youzhi Gu, Fuxu Liu, Xingyu Liao, Shenqi Lai, and Jianyang Gu. A strong baseline and batch normalization neck for deep person re-identification. *IEEE Trans. Multim.*, 22(10):2597–2609, 2020.

Houjing Huang, Dangwei Li, Zhang Zhang, Xiaotang Chen, and Kaiqi Huang. Adversarially occluded samples for person re-identification. In *2018 IEEE Conference on Computer Vision and Pattern Recognition, CVPR 2018, Salt Lake City, UT, USA, June 18-22, 2018*, pages 5098–5107. Computer Vision Foundation / IEEE Computer Society, 2018.

Ilya Loshchilov and Frank Hutter. SGDR: stochastic gradient descent with warm restarts. In *5th International Conference on Learning Representations, ICLR 2017, Toulon, France, April 24-26, 2017, Conference Track Proceedings*. OpenReview.net, 2017.

Jianan Li, Jie Wang, Junjie Chen, and Tingfa Xu. Towards robust point cloud recognition with sample-adaptive auto-augmentation. *IEEE Transactions on Pattern Analysis and Machine Intelligence*, 2025.

Jinxian Liu, Bingbing Ni, Yichao Yan, Peng Zhou, Shuo Cheng, and Jianguo Hu. Pose transferrable person re-identification. In *2018 IEEE Conference on Computer Vision and Pattern Recognition, CVPR 2018, Salt Lake City, UT, USA, June 18-22, 2018*, pages 4099–4108. Computer Vision Foundation / IEEE Computer Society, 2018.

Julia Dietlmeier, Joseph Antony, Kevin McGuinness, and Noel E. O'Connor. How important are faces for person re-identification? In *25th International Conference on Pattern Recognition, ICPR 2020, Virtual Event / Milan, Italy, January 10-15, 2021*, pages 6912–6919. IEEE, 2020.

Kaiyang Zhou, Yongxin Yang, Andrea Cavallaro, and Tao Xiang. Learning generalisable omni-scale representations for person re-identification. *IEEE Trans. Pattern Anal. Mach. Intell.*, 44(9):5056–5069, 2022.

Liang Zheng, Liyue Shen, Lu Tian, Shengjin Wang, Jingdong Wang, and Qi Tian. Scalable person re-identification: A benchmark. In *2015 IEEE International Conference on Computer Vision, ICCV 2015, Santiago, Chile, December 7-13, 2015*, pages 1116–1124. IEEE Computer Society, 2015.

Lin Chen, Hua Yang, Shuang Wu, and Zhiyong Gao. Data generation for improving person re-identification. In Qiong Liu, Rainer Lienhart, Haohong Wang, Sheng-Wei "Kuan-Ta" Chen, Susanne Boll, Yi-Ping Phoebe Chen, Gerald Friedland, Jia Li, and Shuicheng Yan, editors, *Proceedings of the 2017 ACM on Multimedia Conference, MM 2017, Mountain View, CA, USA, October 23-27, 2017*, pages 609–617. ACM, 2017.

Lingxiao He, Xingyu Liao, Wu Liu, Xinchen Liu, Peng Cheng, and Tao Mei. Fastreid: A pytorch toolbox for general instance re-identification. In Abdulmotaleb El-Saddik, Tao Mei, Rita Cucchiara, Marco Bertini, Diana Patricia Tobon Vallejo, Pradeep K. Atrey, and M. Shamim Hossain, editors, *Proceedings of the 31st ACM International Conference on Multimedia, MM 2023, Ottawa, ON, Canada, 29 October 2023- 3 November 2023*, pages 9664–9667. ACM, 2023.

Longhui Wei, Shiliang Zhang, Wen Gao, and Qi Tian. Person transfer GAN to bridge domain gap for person re-identification. In *2018 IEEE Conference on Computer Vision and Pattern Recognition, CVPR 2018, Salt Lake City, UT, USA, June 18-22, 2018*, pages 79–88. Computer Vision Foundation / IEEE Computer Society, 2018.

Ruijie Quan, Xuanyi Dong, Yu Wu, Linchao Zhu, and Yi Yang. Auto-reid: Searching for a part-aware convnet for person re-identification. In *2019 IEEE/CVF International Conference on Computer Vision, ICCV 2019, Seoul, Korea (South), October 27 - November 2, 2019*, pages 3749–3758. IEEE, 2019.

Ryuichiro Hataya, Jan Zdenek, Kazuki Yoshizoe, and Hideki Nakayama. Faster autoaugment: Learning augmentation strategies using backpropagation. In *Computer Vision–ECCV 2020: 16th European Conference, Glasgow, UK, August 23–28, 2020, Proceedings, Part XXV 16*, pages 1–16. Springer, 2020.

Tianlong Chen, Shaojin Ding, Jingyi Xie, Ye Yuan, Wuyang Chen, Yang Yang, Zhou Ren, and Zhangyang Wang. Abd-net: Attentive but diverse person re-identification. In *2019 IEEE/CVF International Conference on Computer Vision, ICCV 2019, Seoul, Korea (South), October 27 - November 2, 2019*, pages 8350–8360. IEEE, 2019.

Wei Li, Rui Zhao, Tong Xiao, and Xiaogang Wang. Deepreid: Deep filter pairing neural network for person re-identification. In *2014 IEEE Conference on Computer Vision and Pattern Recognition, CVPR 2014, Columbus, OH, USA, June 23-28, 2014*, pages 152–159. IEEE Computer Society, 2014.

Xing Fan, Wei Jiang, Hao Luo, and Mengjuan Fei. Spherereid: Deep hypersphere manifold embedding for person re-identification. *J. Vis. Commun. Image Represent.*, 60:51–58, 2019.

Xuelin Qian, Yanwei Fu, Tao Xiang, Wenxuan Wang, Jie Qiu, Yang Wu, Yu-Gang Jiang, and Xiangyang Xue. Pose-normalized image generation for person re-identification. In Vittorio Ferrari, Martial Hebert, Cristian Sminchisescu, and Yair Weiss, editors, *Computer Vision - ECCV 2018 - 15th European Conference, Munich, Germany, September 8-14, 2018, Proceedings, Part IX*, volume 11213 of *Lecture Notes in Computer Science*, pages 661–678. Springer, 2018.

Yifan Sun, Changmao Cheng, Yuhan Zhang, Chi Zhang, Liang Zheng, Zhongdao Wang, and Yichen Wei. Circle loss: A unified perspective of pair similarity optimization. In *2020 IEEE/CVF Conference on Computer Vision and Pattern Recognition, CVPR 2020, Seattle, WA, USA, June 13-19, 2020*, pages 6397–6406. Computer Vision Foundation / IEEE, 2020.

Yixiao Ge, Zhuowan Li, Haiyu Zhao, Guojun Yin, Shuai Yi, Xiaogang Wang, and Hongsheng Li. FD-GAN: pose-guided feature distilling GAN for robust person re-identification. In Samy Bengio, Hanna M. Wallach, Hugo Larochelle, Kristen Grauman, Nicolò Cesa-Bianchi, and Roman Garnett, editors, *Advances in Neural Information Processing Systems 31: Annual Conference on Neural Information Processing Systems 2018, NeurIPS 2018, December 3-8, 2018, Montréal, Canada*, pages 1230–1241, 2018.

Yonggang Li, Guosheng Hu, Yongtao Wang, Timothy Hospedales, Neil M Robertson, and Yongxin Yang. Differentiable automatic data augmentation. In *Computer Vision–ECCV 2020: 16th European Conference, Glasgow, UK, August 23–28, 2020, Proceedings, Part XXII 16*, pages 580–595. Springer, 2020.

Zhedong Zheng, Liang Zheng, and Yi Yang. Unlabeled samples generated by GAN improve the person re-identification baseline in vitro. In *IEEE International Conference on Computer Vision, ICCV 2017, Venice, Italy, October 22-29, 2017*, pages 3774–3782. IEEE Computer Society, 2017.

Zhedong Zheng, Xiaodong Yang, Zhiding Yu, Liang Zheng, Yi Yang, and Jan Kautz. Joint discriminative and generative learning for person re-identification. In *IEEE Conference on Computer Vision and Pattern Recognition, CVPR 2019, Long Beach, CA, USA, June 16-20, 2019*, pages 2138–2147. Computer Vision Foundation / IEEE, 2019.

Zhun Zhong, Liang Zheng, Guoliang Kang, Shaozi Li, and Yi Yang. Random erasing data augmentation. In *Proceedings of the AAAI conference on artificial intelligence*, volume 34, pages 13001–13008, 2020.

Zhongwei Zhao, Ran Song, Qian Zhang, Peng Duan, and Youmei Zhang. Jot-gan: A framework for jointly training GAN and person re-identification model. *ACM Trans. Multim. Comput. Commun. Appl.*, 18(1s):27:1–27:18, 2022.

Zhun Zhong, Liang Zheng, Zhedong Zheng, Shaozi Li, and Yi Yang. Camera style adaptation for person re-identification. In *2018 IEEE Conference on Computer Vision and Pattern Recognition, CVPR 2018, Salt Lake City, UT, USA, June 18-22, 2018*, pages 5157–5166. Computer Vision Foundation / IEEE Computer Society, 2018.

Chapter 3
Auto-Scale Single-Branch Architecture Search for Person Re-identification

3.1 Introduction

Person re-identification (ReID) (Zhou et al. 2019; Wang et al. 2018a; Quan et al. 2019) is a significant research area in computer vision, focused on identifying specific individuals across different camera views or video sequences. Recent advancements in deep learning, particularly with convolutional neural networks (CNNs), have led to remarkable progress in this field. These CNN-based models learn unique features for each input image, aiming to minimize the similarity distance between images of the same identity. ReID presents greater challenges than face recognition. As depicted in Fig. 3.1, hard samples from the Market1501 dataset highlight several difficulties:

- **Color Inconsistency**: Camera variations and different lighting conditions lead to significant color shifts in clothing for the same person.
- **Limited Resolution**: Images are often low-resolution, which hinders the extraction of fine-grained details for identification.
- **Hard Negative Samples**: Similarities between some negative and positive samples can be misleading, making correct identification difficult.

In response to these issues, Zhou et al. (2019) proposed the use of omni-scale features to capture a wide range of discriminatory cues. They define "scale" based on the size of the convolutional kernel, with larger kernels correlating to a broader scale as explained by the effective receptive field theory (Luo et al. 2016). To keep the network architecture compact, they substituted large-scale convolution kernels with a series of smaller 3×3 convolutions, thus introducing the concept of omni-scales. Although OSNet successfully employs this technique, it is not without its drawbacks:

- **Restricted Feature Scale Variety**: The omni-scale block within OSNet only integrates four different scale features, and the fusion of this scale-specific information is accomplished exclusively by the final 1×1 convolution in each block. The ideal

H. Gu et al., *Automated Machine Learning for Person Re-Identification*, Intelligent Perception and Information Processing,
https://doi.org/10.1007/978-981-95-3433-3_3

Fig. 3.1 The problem of person re-identification (ReID) is complex, as shown by the four image sets above. In each group, the first three images depict the same individual, while the last image is of a different person

number of scale features for person re-identification (ReID) tasks has yet to be fully determined.

- **Identical Block Design**: Every block in OSNet utilizes an identical network architecture. This uniform design is suboptimal for CNNs, where a significant difference exists between high-level and low-level features, suggesting that varying architectural configurations might be more appropriate for different stages of the network.

Inspired by neural architecture search (NAS), we introduce an automated scale learning method for person re-identification (ReID) that automatically finds optimal network architectures. This method is centered around our proposed **Auto-Scale Block (ASB)**, which features a bottleneck-like structure reminiscent of ResNet (He et al. 2016). To keep the network lightweight, the ASB uses a combination of 1×1 and efficient 3×3 convolutional operations. The core of the ASB's automatic learning lies in the connections between these 3×3 convolutions. By using NAS to discover the most effective scale combinations, we drastically reduce the need for manual design, leading to a more streamlined architecture.

Our full network, the **Auto-Scale Network (ASNet)**, is built by stacking multiple ASBs. Unlike previous NAS work where blocks often share a common architecture, each ASB within our ASNet is distinct. While this approach offers greater flexibility, it also presents significant search challenges. To address these challenges, we incorporate three types of regularization:

- **Edge Validity Regularization:** Ensures that the connections within the learned architecture are meaningful.
- **Path Validity Regularization:** Guarantees effective information flow through the discovered paths.
- **Graph Consistency Regularization:** Maintains the structural coherence of the network.

These three distinct regularizations are essential for ensuring the effectiveness of the searched architecture. In summary, our main contributions are:

(1) We are the first to apply NAS to discover network architecture topology (not just operations) specifically for the ReID problem.
(2) Our proposed search space allows each block to have a unique architecture, which better aligns with the different feature levels (high vs. low) in CNNs.
(3) We introduce three new regularizations to ensure the validity of edges, paths, and overall graph consistency, which improves the performance of the discovered architectures.
(4) Extensive experiments show that our learned model achieves competitive state-of-the-art results on three ReID benchmarks with a highly efficient parameter count of just 2.2 million.

3.2 Related Works

3.2.1 Person Re-identification

Person re-identification (ReID) is a prominent and challenging research area in computer vision, aiming to retrieve images of a specific person across a network of non-overlapping cameras. Recent years have seen state-of-the-art methods, largely based on deep learning, achieve significant performance gains (Chen et al. 2019a; Zheng et al. 2019a; Wang et al. 2018b; Li et al. 2018; Sun et al. 2018). These approaches can be broadly classified into two main streams based on their feature representation strategy: those utilizing global features and those focusing on local features.

Models based on global features extract a single, holistic feature vector from the entire input image using a backbone network, such as ResNet50 (He et al. 2016) or InceptionV3 (Szegedy et al. 2016). While conceptually simple, these models are often susceptible to challenges like severe occlusions, viewpoint changes, and pedestrian misalignments, which can limit their overall effectiveness and lead to poor performance in real-world scenarios (Sun et al. 2017). In contrast, models based on local features seek to learn more discriminative and robust representations by focusing on distinct regions of a person's body. A seminal work in this domain is the Part-based Convolutional Baseline (PCB) (Sun et al. 2018), which horizontally partitions the convolutional feature map to learn part-specific features, effectively leveraging implicit knowledge of human body structure. The simplicity and efficacy of this strategy have led to its widespread adoption in subsequent research (Zheng et al. 2019b; Wang et al. 2018b; Fu et al. 2019). Notable extensions include the Multiple Granularity Network (MGN) (Wang et al. 2018b), which integrates both global and local features within a multi-branch architecture to achieve a more comprehensive representation. Furthermore, research has delved into more fine-grained methods, such as those that leverage human pose estimation to explicitly handle pose variations and occlusions, thereby improving feature alignment across different camera views.

Beyond manually designed architectures, Neural Architecture Search (NAS) has recently been employed to automatically discover network structures tailored for the person re-identification task (Quan et al. 2019). The approach detailed in this chapter also leverages NAS, but with several key distinctions from prior work. Firstly, our search space, while confined to the connections between pre-defined operations, is more complex and presents a greater search challenge compared to simplified cell-based methods. Secondly, unlike methods such as Auto-ReID (Quan et al. 2019) where Normal and Reduction blocks share a common architecture, our method allows each block to possess a unique, independently searched architecture. This design offers increased flexibility and potential for specialized feature learning that is difficult to achieve through a shared-cell paradigm.

3.2.2 Neural Architecture Search

Neural Architecture Search (NAS) seeks to automate the design of high-performance neural network architectures, thereby reducing the need for manual intervention and domain expertise. Early NAS methods, often relying on reinforcement learning (Zoph and Le 2017; Zoph et al. 2018) or evolutionary algorithms, were computationally prohibitive, frequently requiring thousands of GPU-days to find a suitable architecture. These methods typically operated by training and evaluating each candidate architecture from scratch, often on a small proxy dataset (e.g., CIFAR-10) before transferring it to a larger target dataset, a process that was both time-consuming and prone to an "architecture-transfer" performance gap.

To mitigate the substantial computational cost, more efficient methods such as one-shot (Brock et al. 2018; Bender et al. 2018) and differentiable (Liu et al. 2019) NAS were introduced. Differentiable Architecture Search (DARTS) (Liu et al. 2019) pioneered the relaxation of the discrete architectural search space into a continuous one, enabling efficient architecture optimization via gradient descent. This breakthrough significantly reduced search times to just a few GPU-days, making it feasible to perform the search directly on large-scale target datasets, thus eliminating the proxy paradigm. However, this paradigm is not without its challenges. Subsequent research has revealed that DARTS can suffer from instability, a documented tendency to converge to simple architectures with an excessive number of skip-connections, and a noticeable performance gap between the continuously optimized supernet and the final discrete architecture.

The methodology presented in this chapter is founded on differentiable NAS principles, yet it is specifically designed to overcome the documented limitations of DARTS. Our approach imposes fewer manual constraints on the cell architecture; for example, it does not limit the aggregation of feature maps to inputs from only two preceding operations. Furthermore, by focusing the search exclusively on the connectivity patterns between a fixed set of operations, our method does not search for the operation types themselves. This targeted search strategy avoids the instability and the documented preference for skip-connections observed in DARTS, leading to the discovery of more robust and specialized architectures for person re-identification.

3.3 Methodology

In this section, we first specify the fundamental building block, termed the auto-scale block, in Sect. 3.3.1. This block is the core component from which the entire auto-scale network, detailed in Sect. 3.3.2, is constructed. The network learns multi-scale features tailored for the ReID task by optimizing the connections between lightweight 3×3 operations within each block. To achieve this, we employ a differentiable neural architecture search method, as described in Sect. 3.3.3. Furthermore, to address challenges encountered during the search process and enhance the final model's performance, we implement three distinct regularization techniques, which are presented in Sect. 3.3.4.

3.3.1 Auto-Scale Block

3.3.1.1 Node Operations

The primary function of nodes in the Auto-Scale Block (ASB) is feature processing. As illustrated in Fig. 3.2a, the fundamental process involves an aggregation step, where features from preceding nodes are combined to form a single aggregated feature. This aggregated feature is then processed by a specific "operation" resulting in the final processed feature, which is subsequently distributed to succeeding nodes. The focus of this chapter is on exploring the network's wiring paths, therefore the specific types of operations used are not the main subject of our investigation. For creating a lightweight network architecture, the ASB is primarily constructed using the Lite 3×3 and RCB 1×1 operations. The detailed structure of these two operations is provided in Fig. 3.2.

As shown in Fig. 3.2, the operations we employ utilize a ReLU-Conv-BN triplet, which differs from the conventional Conv-BN-ReLU sequence. While these two sequences are functionally equivalent when a network operation has only a single preceding operation, our approach involves feature aggregation. The use of the ReLU-Conv-BN triplet allows it to receive both positive and negative features from preceding operations, which is crucial for preventing feature inflation when a node has a high input degree.

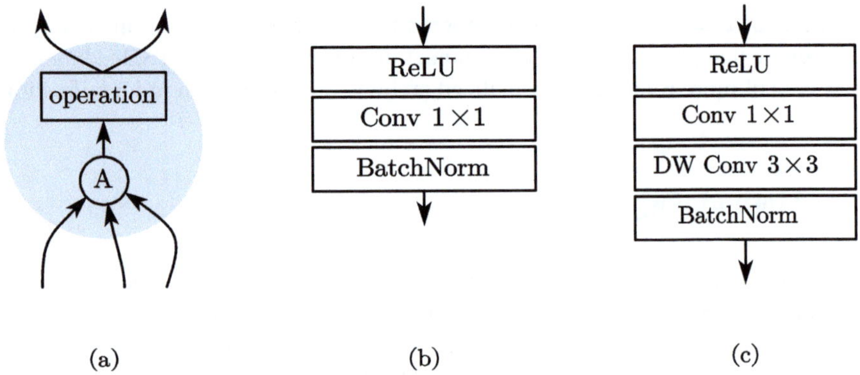

Fig. 3.2 Node Operations. **a** Shows the aggregation operation, denoted by "A", which combines features from different sources. The specific "operation" detailed in (**a**) corresponds to either the RCB 1×1 operation **b** or the Lite 3×3 operation (**c**), both of which are core components of our architecture. DW denotes the depthwise convolution

3.3.1.2 Edge Operations

The edges within the ASB solely serve as data conduits, enabling the transfer of features from one node to another.

3.3.1.3 The Overall Structure

Following the definitions of nodes and edges, the concrete architecture of the ASB can be formally specified. The input feature map, denoted as $\mathbf{X}_{in} \in \mathbb{R}^{B \times C \times H \times W}$, is initially processed by an RCB 1×1 operation (indexed as 0). This operation yields the feature map $\mathbf{X}_0 \in \mathbb{R}^{B \times C' \times H \times W}$, which has a reduced dimensionality where $C' = C/r$.

$$\mathbf{X}_0 = RCB\left(\mathbf{X}_{in}\right), \tag{3.1}$$

where $RCB(\cdot)$ signifies the RCB 1×1 operation.

Next, we define N Lite 3×3 nodes, indexed sequentially from $1, 2, ..., N$. The output feature map $\mathbf{X}_j \in \mathbb{R}^{B \times C' \times H \times W}$ for any given node j is defined by the following expression:

$$\mathbf{X}_j = Lite\left(\sum_{i=0}^{j-1} \mathbf{E}_{i,j}\mathbf{X}_i\right), \tag{3.2}$$

where $Lite(\cdot)$ denotes the Lite 3×3 operation. The binary variable $\mathbf{E}_{i,j} \in \{0, 1\}$ indicates the presence or absence of a connection from node i to node j. Specifically, $\mathbf{E}_{i,j} = 1$ signifies the existence of an edge, whereas $\mathbf{E}_{i,j} = 0$ indicates its absence.

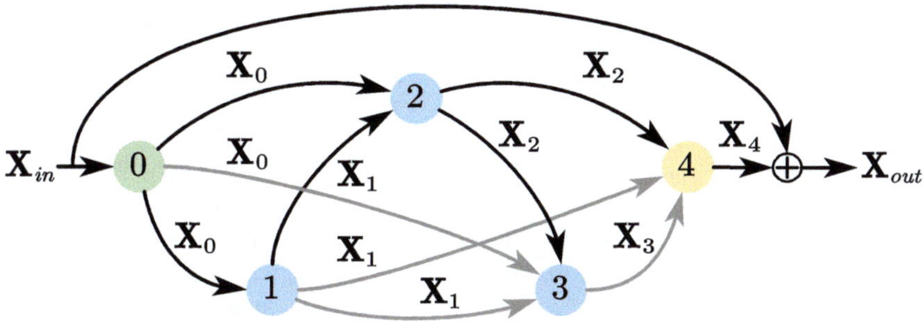

Fig. 3.3 Auto-Scale Block for $N = 3$. Node 0 and Node 4 are configured with the RCB 1×1 operation, while Nodes 1, 2, and 3 employ the Lite 3×3 operation. The network connectivity is indicated by the lines: gray lines signify non-existent connections (e.g., $\mathbf{E}_{0,3} = 0$), whereas black lines denote active, existing connections (e.g., $\mathbf{E}_{0,1} = 1$)

As is evident from Eq. (3.2), the aggregation mechanism for the Lite 3×3 is a direct summation of the features from the preceding nodes. This approach offers the primary benefit of preserving the channel count, unlike concatenation operations, while also contributing to a significantly smaller overall network parameter count.

Finally, an additional RCB 1×1 operation (indexed $N + 1$) is required to restore the dimensionality of the feature map to its original size, resulting in $\mathbf{X}_{N+1} \in \mathbb{R}^{B \times C \times H \times W}$.

$$\mathbf{X}_{N+1} = RCB\left(\sum_{i=1}^{N} \mathbf{E}_{i,N+1} AG(\mathbf{X}_i)\right), \tag{3.3}$$

where $AG(\cdot)$ denotes the aggregation gate (Zhou et al. 2019). This specific aggregation operation is applied to the RCB 1×1 and is shared across all feature streams within the same ASB. To further stabilize the optimization process, we incorporate a residual connection (He et al. 2016) into our ASB. Consequently, the final output feature map, $\mathbf{X}_{out} \in \mathbb{R}^{B \times C \times H \times W}$, is defined as:

$$\mathbf{X}_{out} = \mathbf{X}_{in} + \mathbf{X}_{N+1}. \tag{3.4}$$

To ensure valid edge connections, a set of constraints is imposed on the matrix $\mathbf{E} \in \mathbb{R}^{(N+2)\times(N+2)}$:

(1) $\mathbf{E}_{0,N+1} = 0$. No edge exists from node 0 to node $N + 1$.
(2) $\mathbf{E}_{i,j} = 0, \forall i \geq j$. Edges must only flow from a node with a smaller index to one with a larger index.

Consequently, the maximum number of edges is limited to $(N^2 + N)/2 + N$. An example of the ASB architecture for $N = 3$ is illustrated in Fig. 3.3. We designate this set of all valid edges as $\mathcal{V} = \{(i, j)\}$. The primary objective of our work, as detailed in Sect. 3.3.3, is to determine the optimal configuration of the binary variables

Table 3.1 Architecture of the ASNet with a 256 × 128 input image resolution. This configuration is based on a design where $N = 10$

Layer name	Output size	ASNet
conv1	128×64, 64	7×7 conv, stride 2
	64×32, 64	3×3 max pool, stride 2
Conv2_x	64×32, 256	ASB×2
Transition	64×32, 256	1×1 conv
	32×16, 256	2×2 average pool, stride 2
Conv3_x	32×16, 384	ASB×2
Transition	32×16, 384	1×1 conv
	16×8, 384	2×2 average pool, stride 2
Conv4_x	16×8, 512	ASB×2
Conv5	16×8, 512	1×1 conv
Gap	1×1, 512	Global average pool
# params		2.2M

$\{\mathbf{E}_{i,j} | (i, j) \in \mathcal{V}\}$ to create an architecture capable of learning auto-scale features. It is noteworthy that the OSB (Zhou et al. 2019) can be considered a specific instance of our ASB. The key distinction lies in the fact that OSB is based on a manually crafted design, whereas our ASB is derived through an automated process.

3.3.2 *Auto-Scale Network*

The Auto-Scale Network (ASNet) architecture is built upon a sequence of stacked ASBs. The complete network structure is detailed in Table 3.1, which shows that ASNet is composed of six ASBs. It is important to note that, in contrast to DARTS (Liu et al. 2019), where architectural cells are shared across stages (e.g., Normal and Reduction Cells), the architecture of each individual ASB within our ASNet is unique and not shared. This distinction leads to a significantly larger and more intricate search space. Consequently, the size of the total search space is estimated to be approximately 2^{3N^2+9N}.[1]

As indicated in the table, a transition layer is also incorporated into the architecture. This layer's purpose is to downsample the feature map's height and width by a factor of two. It is implemented as a combination of a 1 × 1 convolution and a 2 × 2 average pooling operation.

Beyond the ASNet backbone architecture, a classification layer is also required. We employ BNNeck (Luo et al. 2019) for this purpose, a method widely used in the field of person re-identification. More specifically, for an input image $\mathbf{I}_i \in \mathbb{R}^{3\times H\times W}$

[1] In this chapter, with $N = 10$, there are approximately 2.52×10^{117} possible architectures for ASNet.

with its corresponding label $y_i \in \{1, 2, \ldots, N_c\}$, the ASNet produces a feature vector $f_i \in \mathbb{R}^{512}$. Following the BNNeck approach, the triplet loss is then applied to f_i as follows:

$$\mathcal{L}_{triplet} = \frac{1}{PK} \sum_{i=1}^{PK} \left[m + \max_{y_j = y_i} \| f_i - f_j \| - \min_{y_j \neq y_i} \| f_i - f_j \| \right]_+ . \tag{3.5}$$

The mini-batch is formed using PK sampling (Luo et al. 2019), consisting of P distinct identities and K images for each identity. Subsequently, the ID loss is computed using the feature $f_i^{(bn)}$, which is derived by applying a batch normalization to f_i. The ID loss is formally defined as:

$$\mathcal{L}_{ID} = -\frac{1}{N} \sum_{i=1}^{N} \log \frac{\exp\left(w_{y_i}^T f_i^{(bn)}\right)}{\sum_{c=1}^{N_c} \exp\left(w_c^T f_i^{(bn)}\right)}, \tag{3.6}$$

where w_c is the weight vector for class c. Thus, the total loss function for the entire network is formulated as:

$$\mathcal{L} = \mathcal{L}_{ID} + \mathcal{L}_{triplet}. \tag{3.7}$$

3.3.3 Search Strategy

This section details our approach for optimizing the architectural parameters, represented by $\mathcal{E} = \{\mathbf{E}_{i,j}^b | (i, j) \in \mathcal{V}, b = 1, 2, \ldots, 6\}$, across the six ASBs within ASNet. Given a training dataset $\mathcal{D}_{trn}$ and a validation dataset $\mathcal{D}_{val}$, our optimization problem is formulated as:

$$\begin{aligned} \mathcal{E}^\star &= \arg\min_{\mathcal{E}} \mathcal{L}\left(\mathcal{D}_{val}; W^\star(\mathcal{E}), \mathcal{E}\right), \\ \text{s.t. } W^\star(\mathcal{E}) &= \arg\min_{W} \mathcal{L}\left(\mathcal{D}_{trn}; W, \mathcal{E}\right), \end{aligned} \tag{3.8}$$

where $\mathcal{L}(\mathcal{D}; W, \mathcal{E})$ denotes the loss from Eq. (3.7) when the network parameters are W and the architectural parameters are $\mathcal{E}$. This framework constitutes a bilevel optimization problem.

Finding the optimal $\mathcal{E}^\star$ is a challenging task, particularly because the architectural parameters $\mathbf{E}_{i,j}^b \in \{0, 1\}$ are binary, which significantly complicates the optimization process. To enable a continuous search space, we relax the binary path selection by applying a sigmoid function to a continuous variable:

$$\begin{aligned} \tilde{\mathbf{E}}_{i,j}^b &= \sigma\left(\alpha_{i,j}^b\right) \\ &= \frac{1}{1 + \exp\left(-\alpha_{i,j}^b\right)}, \end{aligned} \tag{3.9}$$

where $\alpha_{i,j}^b$ is the continuous counterpart to the binary variable $\mathbf{E}_{i,j}^b$. This relaxation transforms our optimization task into the search for the optimal continuous variables $\tilde{\mathcal{E}} = \{\alpha_{i,j}^b | (i, j) \in \mathcal{V}, b = 1, 2, \ldots, 6\}$.

$$\begin{aligned} \tilde{\mathcal{E}}^\star &= \arg\min_{\tilde{\mathcal{E}}} \mathcal{L}\left(\mathcal{D}_{val}; W^\star\left(\tilde{\mathcal{E}}\right), \tilde{\mathcal{E}}\right), \\ \text{s.t. } W^\star\left(\tilde{\mathcal{E}}\right) &= \arg\min_{W} \mathcal{L}\left(\mathcal{D}_{trn}; W, \tilde{\mathcal{E}}\right). \end{aligned} \tag{3.10}$$

This reformulated optimization problem can be solved using an approach similar to DARTS. Specifically, in each training iteration, the network parameters W and the relaxed architectural parameters $\tilde{\mathcal{E}}$ are optimized alternately. Upon convergence, the final discrete architecture is derived by applying a thresholding step: we set $\mathbf{E}_{i,j}^b$ to 1 if the corresponding $\tilde{\mathbf{E}}_{i,j}^b$ exceeds a pre-defined threshold τ, and to 0 otherwise.

3.3.4 Regularizations

3.3.4.1 Entropy Regularization

When the objective is directly optimized according to Eq. (3.10), the parameters $\{\tilde{\mathbf{E}}_{i,j}^b | (i, j) \in \mathcal{V}, b = 1, 2, \ldots, 6\}$ are typically distributed continuously between 0 and 1, as illustrated in Fig. 3.4a. This continuous nature leads to a significant discretization gap when applying a pre-defined threshold τ to derive the final network architecture, a challenge also noted in methods like DARTS (Liu et al. 2019). To mitigate this gap, we introduce an additional regularization term that encourages the values of $\{\tilde{\mathbf{E}}_{i,j}^b | (i, j) \in \mathcal{V}, b = 1, 2, \ldots, 6\}$ to converge towards either 0 or 1. This regularization is directly defined by the entropies of these parameters.

$$\begin{aligned} \mathcal{L}_e = &- \frac{1}{6|\mathcal{V}|} \sum_{b=1}^{6} \sum_{(i,j)\in\mathcal{V}} \sigma\left(\alpha_{i,j}^b\right) \log\left(\sigma\left(\alpha_{i,j}^b\right)\right) \\ &+ \left(1 - \sigma\left(\alpha_{i,j}^b\right)\right) \log\left(1 - \sigma\left(\alpha_{i,j}^b\right)\right). \end{aligned} \tag{3.11}$$

In this way, the optimization problem of Eq. (3.10) can be changed to

$$\begin{aligned} \tilde{\mathcal{E}}^\star &= \arg\min_{\tilde{\mathcal{E}}} \mathcal{L}\left(\mathcal{D}_{val}; W^\star\left(\tilde{\mathcal{E}}\right), \tilde{\mathcal{E}}\right) + \lambda_e \mathcal{L}_e, \\ \text{s.t. } W^\star\left(\tilde{\mathcal{E}}\right) &= \arg\min_{W} \mathcal{L}\left(\mathcal{D}_{trn}; W, \tilde{\mathcal{E}}\right). \end{aligned} \tag{3.12}$$

Figure 3.4b depicts the impact of incorporating entropy regularization on the parameters $\{\tilde{\mathbf{E}}_{i,j}^1 | (i, j) \in \mathcal{V}\}$. It is evident that this technique drives the values of these parameters towards either 0 or 1, thereby creating a more binary distribution. This

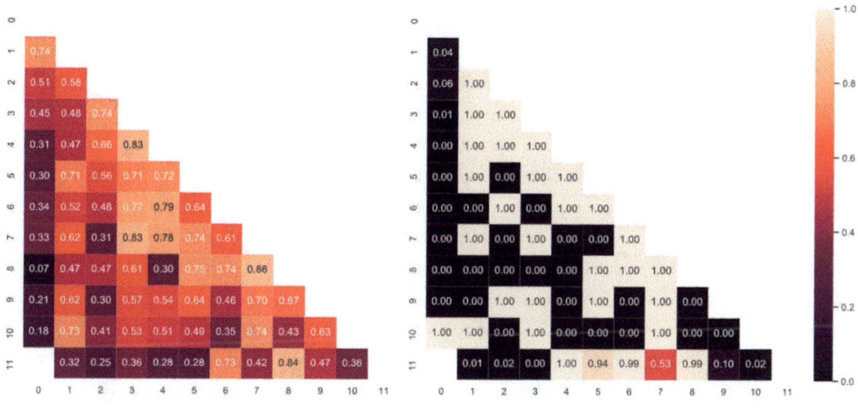

(a) (b)

Fig. 3.4 Visualization of the architecture parameters $\{\tilde{\mathbf{E}}^1_{i,j}\}$ for a single cell. **a** illustrates the parameters without the application of entropy regularization. In contrast, **b** shows the effect of applying entropy regularization, specifically with a value of $\lambda_e = 5$

effect substantially mitigates the error introduced during the binarization step of the network search process.

3.3.4.2 Validity Regularization

As indicated in Sect. 3.1, our six ASBs are constructed independently, which poses a potential issue. Specifically, blocks located closer to the network input may become overly simplistic. With the current threshold τ, it is conceivable that no valid path can be formed from node 0 to node $N+1$. This is primarily due to the fact that the gradients for the initial blocks are often smaller, leading to a slower update rate compared to subsequent blocks. Even with the incorporation of residual connections, this problem is not entirely alleviated. Although a valid path could be guaranteed by lowering the threshold τ, this action would exacerbate the discretization gap problem mentioned earlier. Consequently, to guarantee the existence of a valid path from node 0 to node $N+1$ in the final architecture, we propose a novel validity regularization term.

To formalize the concept of a valid path, we define F_i as the maximum amount of information that can be transmitted from node 0 to node i, with $F_0 = 1$. The parameter $\tilde{\mathbf{E}}_{i,j}$ can then be interpreted as the efficiency of information transmission from node i to node j. We can update the amount of information each node receives from node 0 using the following recursive formula:

$$F_i = \begin{cases} 1 & i = 0 \\ \max\limits_{(j,i)\in\mathcal{V}} F_j \tilde{\mathbf{E}}_{j,i} & i > 0 \end{cases} \tag{3.13}$$

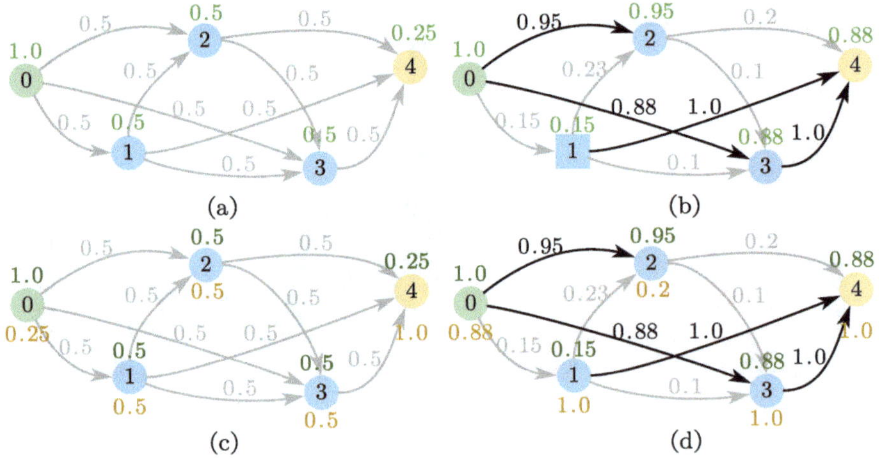

Fig. 3.5 Visualization of the Validity and Consistent Regularization. The figure omits the residual connection for clarity and uses a threshold $\tau = 0.85$. The green numbers represent the forward path values $\{F_i\}$, while the yellow numbers represent the backward path values $\{B_i\}$, both for nodes $i = 0, \ldots, N+1$. **a** Displays the initial values of $\{F_i\}$. **b** Shows the values of $\{F_i\}$ during the training phase. **c** Provides the initial values of $\{B_i\}$. **d** Illustrates the values of $\{B_i\}$ during the training process

In order to ensure that a valid path always exists from node 0 to node $N+1$, we define the validity regularization term as:

$$\mathcal{L}_v = \frac{1}{6}\sum_{b=1}^{6}\left(F_{N+1}^b - 1\right)^2. \tag{3.14}$$

As shown in Fig. 3.5a, an initial F_{N+1} value of 0.25 clearly indicates a lack of a valid path. In contrast, Fig. 3.5b depicts an F_{N+1} of 0.88, confirming the presence of a valid path from node 0 to node $N+1$. This demonstrates that a higher value of F_{N+1} directly correlates with a greater likelihood of a valid path.

3.3.4.3 Consistent Regularization

Even with the introduction of validity regularization, a discontinuity problem can persist. For instance, as shown in Fig. 3.5b, the edge from node 0 to node 2 is non-essential, as it does not contribute to a valid path to the output node. Such superfluous edges also contribute to the discretization gap. The independent definition of the parameters $\{\tilde{\mathbf{E}}_{i,j}^b | (i,j) \in \mathcal{V}, b = 1, 2, \ldots, 6\}$ based on their corresponding architectural parameters $\{\alpha_{i,j}^b | (i,j) \in \mathcal{V}, b = 1, 2, \ldots, 6\}$ means there is an inherent lack of correlation between the edges. To address this and reduce the number of redundant edges, we introduce a consistent regularization term.

We define a reverse information flow, where B_i represents the amount of information transmittable from node $N+1$ to node i in a reverse-gated graph, with $B_{N+1} = 1$:

$$B_i = \begin{cases} 1 & i = N+1 \\ \max\limits_{(i,j)\in\mathcal{V}} B_j \tilde{\mathbf{E}}_{i,j} & i < N+1 \end{cases} \tag{3.15}$$

The key distinction between B_i and the previously defined forward information flow F_i is that B_i is computed via backward propagation. Our consistent regularization is designed to align these forward and backward information flows:

$$\mathcal{L}_c = \frac{1}{6N}\sum_{b=1}^{6}\sum_{i=1}^{N}\left(F_i^b - B_i^b\right)^2. \tag{3.16}$$

Figure 3.5d illustrates this concept. A scenario where $F_2 = 0.95$ and $B_2 = 0.2$ results in a large penalty value according to Eq. (3.16). This penalty encourages two outcomes: either a reduction in F_2 to eliminate the redundant edge from node 0 to node 2, or an increase in B_2 by introducing a new edge from node 2 to node 4 to balance the flow.

Algorithm 1 Auto-Scale Learning for ReID

Require: The architecture parameter $\tilde{\mathcal{E}}$, the network parameter W, the training set $\mathcal{D}_T$, and the threshold τ.
Ensure: The optimal architecture $\mathcal{E}^\star$.
1: Split $\mathcal{D}_T$ into a search training set $\mathcal{D}_{trn}$ and a search validation set $\mathcal{D}_{val}$.
2: **while** not converged **do**
3: Get batch data d_{trn} from $\mathcal{D}_{trn}$.
4: Update W by minimizing the loss $\mathcal{L}(d_{trn}; W, \tilde{\mathcal{E}})$.
5: Get batch data d_{val} from $\mathcal{D}_{val}$.
6: Update $\tilde{\mathcal{E}}$ by minimizing the loss $\mathcal{L}(d_{val}; W, \tilde{\mathcal{E}}) + \lambda_e\mathcal{L}_e + \lambda_v\mathcal{L}_v + \lambda_c\mathcal{L}_c$.
7: **end while**
8: Derive the optimal architecture $\mathcal{E}^\star$ from $\tilde{\mathcal{E}}$ using the threshold τ.

By incorporating all three regularization terms, the final optimization problem is defined as:

$$\begin{aligned} \tilde{\mathcal{E}}^\star &= \arg\min_{\tilde{\mathcal{E}}} \mathcal{L}\left(\mathcal{D}_{val}; W^\star\left(\tilde{\mathcal{E}}\right), \tilde{\mathcal{E}}\right) + \lambda_e\mathcal{L}_e \\ &+ \lambda_v\mathcal{L}_v + \lambda_c\mathcal{L}_c, \\ \text{s.t.}\, W^\star\left(\tilde{\mathcal{E}}\right) &= \arg\min_{W} \mathcal{L}\left(\mathcal{D}_{trn}; W, \tilde{\mathcal{E}}\right), \end{aligned} \tag{3.17}$$

where λ_e, λ_v, and λ_c are hyperparameters that control the weighting of each regularization term. The complete procedure of our auto-scale learning for person re-identification is presented in Algorithm 1.

3.3.4.4 Discussion

In this subsection, we have introduced three distinct regularization techniques designed to ensure the validity of edges, the existence of valid paths, and the overall consistency of the graph structure. Each of these regularizations serves a unique and indispensable purpose. The entropy regularization's primary role is to minimize the discretization gap, and its influence is continuous throughout the entire training process. Validity regularization guarantees the existence of a valid path from node 0 to node $N + 1$, a critical function that is most impactful during the early stages of training. Finally, consistent regularization refines the searched architecture by eliminating superfluous or meaningless edges, with its effect becoming more pronounced in the middle and later stages of training.

3.4 Experiments

3.4.1 Datasets and Evaluation Metrics

Datasets. The performance of our proposed model is evaluated on three widely used person re-identification datasets: Market1501 (Zheng et al. 2015), CUHK03 (Li et al. 2014), and MSMT17 (Wei et al. 2018).

- **Market1501** (Zheng et al. 2015) is a large-scale person ReID dataset collected from six cameras. It comprises 12,396 training images of 751 distinct identities and a test set with 19,372 gallery images and 3,368 query images, covering 750 identities. The training and test sets have no identity overlap, and each identity in the training set has an average of 17.2 images.
- **CUHK03** (Li et al. 2014) consists of 1,467 identities and 28,192 bounding boxes. The dataset is split into 26,264 training images for 1,367 identities and 1,928 test images for 100 identities. Following the new evaluation protocol proposed by (Zhong et al. 2017), we use this split for our experiments. The dataset provides two distinct labeling methods: manual annotations and automatically detected bounding boxes.
- **MSMT17** (Wei et al. 2018) is currently the largest-scale person ReID dataset, featuring 126,441 images of 4,101 identities captured by 15 cameras. The training set includes 32,621 bounding boxes from 1,041 identities, while the test set comprises 93,820 bounding boxes from 3,060 identities. From the test set, 11,659 images serve as queries, with the other 82,161 bounding boxes used for the gallery. This benchmark is particularly challenging due to its diverse and complex scenes, encompassing both indoor and outdoor environments.

Evaluation Metrics. To evaluate the performance of our approach and enable a comparison with state-of-the-art methods, we report two standard evaluation metrics. Following common practices (Zheng et al. 2015; Wei et al. 2018), we present the cumulative matching characteristics (CMC) at rank-1 and the mean average precision (mAP) on the aforementioned three benchmarks.

3.4.2 Implementation Details

3.4.2.1 Searching Configuration

For the search phase, we split the official training set into two subsets: 50% of the images are randomly selected to form the search training set ($\mathcal{D}_{trn}$), with the remaining images constituting the search validation set ($\mathcal{D}_{val}$). We perform the architecture search for our ASNet on the Market1501 dataset and then transfer the resulting architecture to other datasets, a strategy similar to Auto-ReID. To maintain parameter consistency with OSNet, the number of operations, N, within each block is set to 10. The weighting coefficients for the regularizations, λ_e, λ_v, and λ_c, are set to 5, 1, and 2, respectively, unless stated otherwise. Input images are resized to 256×128. For the triplet loss, the margin m is set to 0.3, and we use a batch sampling strategy of $P = 16$ identities with $K = 4$ images per identity. The network is optimized using SGD for 350 epochs. The initial learning rate is 0.065 and is reduced by a factor of 0.1 at epochs 150, 225, and 300. The final searched model is obtained by applying a threshold τ of 0.85. The entire search process takes approximately 10 h on a single NVIDIA Tesla V100 GPU.

3.4.2.2 Training Configuration

Once the final searched model is obtained, it can be trained from scratch using the same configuration as the search phase. Alternatively, for fine-tuning a model pre-trained on ImageNet, we use the AMSGrad optimizer with an initial learning rate of 0.0015. The model is trained for 250 epochs, with the learning rate decaying according to a cosine annealing strategy (without restarts). In the initial 10 epochs, the pre-trained backbone is frozen, and only the randomly initialized classifier is trained.

3.4.3 Ablation Study

3.4.3.1 The Variants of the Omni-Scale Network

Given that our search space is primarily derived from OSNet, we begin our ablation studies by comparing several of its variants, as detailed in Table 3.2. While triplet loss is known to produce more compact feature representations in many current works, the

Table 3.2 The mAP and Rank-1 on the Market1501 dataset for various OSNet configurations. † denotes results originally cited from (Zhou et al. 2019). OSNet‡ refers to the standard OSNet architecture as described in Zhou et al. (2019) but with the final fully connected layer removed and without the aggregation gate defined in Eq.(3). OSNet∗ indicates the original OSNet with the last fully connected layer excluded. "BNNeck" represents the classification layer proposed in Luo et al. (2019). "Tri" signifies the use of triplet loss during training

Model	Architecture	mAP	R1
1	OSNet†	81.0	93.6
2	OSNet(Our Imple.)	82.15	93.35
3	OSNet+Tri	–	–
4	OSNet+BNNeck+Tri	–	–
5	OSNet‡+BNNeck+Tri	83.34	93.43
6	OSNet∗+BNNeck+Tri	**84.16**	**93.89**

original OSNet implementation did not include it. We therefore sought to integrate triplet loss into the OSNet architecture.

A direct addition of triplet loss to the OSNet model led to a failure to converge. As shown by Model 3 and Model 4 in Table 3.2, this issue persisted even with the inclusion of a BNNeck (Luo et al. 2019) layer. Through multiple experiments, we discovered that removing the final fully connected (fc) layer from the original OSNet and replacing it with BNNeck significantly improved the model's performance, as evidenced by Model 6. Furthermore, a comparison between Model 5 and Model 6 demonstrates that the aggregation gate contributes to better model performance. Following this comparative analysis of OSNet variants, we established Model 6 as our baseline. Our subsequent search space was then designed based on this refined architecture.

3.4.3.2 The Effectiveness of the Regularizations

We begin by examining the effectiveness of entropy regularization. To demonstrate the existence of the discretization gap, we compare the performance of sub-networks obtained with different thresholds τ against the performance of the hyper-network. It is important to note that the weights for both the hyper-network and the sub-networks are inherited from the trained hyper-network weights W, which are obtained following the procedure in Algorithm 1.

As shown in Table 3.3, the performance of the hyper-network without entropy regularization is superior to that of the hyper-network with the regularization. This marginal drop in mAP with the addition of entropy regularization is expected, as a regularization term introduces an auxiliary loss that may not align perfectly with the primary objective function, thus slightly diminishing the overall performance on the hyper-network. However, without entropy regularization, directly deriving

Table 3.3 The mAP and Rank-1 accuracy on the Market1501 dataset, comparing results with and without entropy regularization. "ER" denotes entropy regularization. "H" refers to the hyper-network used during the search phase, where the sub-network's architecture is determined by the current τ, and its parameter weights are directly inherited from the hyper-network. "TS" indicates training from scratch on the optimal sub-network architecture, using $\tau = 0.4$ for the model without ER and $\tau = 0.9$ for the model with ER

		w/o ER		w ER	
		mAP	R1	mAP	R1
H		75.17	89.19	74.64	88.98
τ	0.0	0.36	0.77	11.24	29.57
	0.1	0.36	0.68	64.14	81.38
	0.2	0.38	0.68	64.14	81.38
	0.3	0.69	1.81	68.81	84.89
	0.4	3.04	9.86	68.81	84.89
	0.5	–	–	68.81	84.89
	0.6	–	–	68.81	84.89
	0.7	–	–	68.81	84.89
	0.8	–	–	68.81	84.89
	0.9	–	–	68.81	84.89
	1.0	–	–	–	–
TS		73.66	88.34	75.14	89.23

a sub-network from the hyper-network using a threshold τ results in a significant performance degradation, highlighting a substantial discretization gap.

With entropy regularization, the performance difference between the hyper-network and the best sub-network is much smaller, with only an approximately 6% reduction in mAP and 4% in Rank-1. This result is a clear improvement over the non-regularized case and confirms that entropy regularization effectively reduces the discretization gap. Furthermore, when we train the best sub-networks (found in both the regularized and non-regularized cases) from scratch, we observe that the sub-network discovered with entropy regularization consistently outperforms its counterpart. In summary, the application of entropy regularization is crucial for our approach.

Next, we explain the impact of validity regularization. The search results for the 'conv2_1' block without validity regularization are shown in Fig. 3.6a, d. From these figures, it is evident that at a threshold of $\tau = 0.85$, no valid path exists from node 0 to node 11. This is because the corresponding parameter, $\tilde{\mathbf{E}}^1_{0,1}$, converges to a value of 0.4. This outcome is attributed to the search process itself: at later stages, sharp adjustments to the $\{\tilde{\mathbf{E}}_{i,j} | (i, j) \in \mathcal{V}\}$ values can introduce significant fluctuations in the overall objective function. Consequently, some of these parameters converge to values that are neither close to 0 nor 1. This phenomenon makes it highly probable that a valid path will not exist, regardless of the chosen value for λ_e.

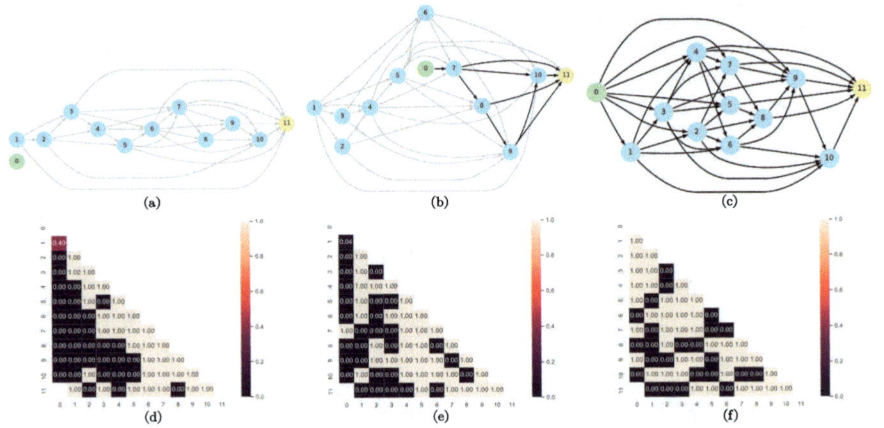

Fig. 3.6 Visualization of the conv2_1 blocks obtained using different regularization schemes. **a** shows the block searched with Entropy Regularization (ER). **b** shows the block searched with both ER and Validity Regularization (VR). **c** shows the block searched with ER, VR, and Consistent Regularization (CR). **d**, **e**, and **f** display the corresponding final architecture parameters $\{\tilde{\mathbf{E}}^{1}_{i,j}|(i,j) \in \mathcal{V}\}$ for each of these three search processes, respectively. A threshold of $\tau = 0.85$ was applied in all cases

Table 3.4 The mAP and Rank-1 accuracy on the Market1501 dataset for both manually designed models and models obtained through architectural search with different regularization techniques. "ER" denotes entropy regularization, "VR" denotes validity regularization, and "CR" denotes consistent regularization

Model	mAP	R1
OSNet+BNNeck+Tri	84.16	93.89
Connect-All	84.23	93.67
Baseline	83.15±0.58	93.22±0.21
Baseline+ER	84.36±0.43	93.87±0.16
Baseline+ER+VR	84.75±0.33	93.97±0.23
Baseline+ER+VR+CR	**85.11±0.12**	**94.23±0.19**

However, the introduction of validity regularization significantly reduces the probability of this issue occurring. The search results for the same block with this regularization are presented in Fig. 3.6b, e. While the validity regularization ensures a path, the resulting architecture still contains numerous meaningless edges, which also contribute to the discretization gap.

When we further add consistent regularization, the final searched architecture is as shown in Fig. 3.6c, f. A comparison across all parts of Fig. 3.6 demonstrates that the combined application of all three regularizations—entropy, validity, and consistency—effectively ensures a well-formed architecture by guaranteeing valid edges, complete paths, and overall graph consistency.

As shown in Table 3.4, we compare the performance of manually designed models against our searched models with different regularizations on the Market1501 dataset. The first two rows represent hand-designed architectures: "OSNet+BNNeck+Tri" is the model 6 from Table 3.2, while "Connect-All" represents a fully connected architecture where all parameters $\{\mathbf{E}_{i,j}^{b} | (i, j) \in \mathcal{V}, b = 1, \ldots, 6\}$ are set to 1. The last four rows are models discovered by our search method under various conditions, with each experiment conducted four times with different random seeds. "Baseline" refers to a model searched without any regularization, and the threshold τ is chosen using the same methodology as in Table 3.3. The threshold τ for the subsequent three models with regularization is fixed at 0.85.

A comparison of the results in Table 3.4 reveals that there is a performance gap between the two hand-designed architectures and our final searched model. Among the three regularization methods proposed, Entropy Regularization (ER) is most critical as it first addresses the discretization gap, leading to the largest performance gain over the "Baseline". The subsequent introduction of Validity Regularization (VR) and Consistent Regularization (CR) further enhances the stability and effectiveness of the searched architecture by enforcing valid paths and graph consistency, respectively.

3.4.3.3 The Visualization of the Searched Architecture

The best network architecture we discovered is visualized in Fig. 3.7. Unlike OSNet, the architectures of the six blocks in our searched network are all different, a complexity that is exceedingly difficult to achieve through manual design. While our blocks contain the same number of Lite 3×3 operations as OSNet, they are capable of expressing features with a much richer variety of scales. For example, considering "conv2_1" in Fig. 3.7, the network can represent seven different types of scales through various paths, such as $0 \rightarrow 9 \rightarrow 11$, $0 \rightarrow 7 \rightarrow 9 \rightarrow 11$, $0 \rightarrow 4 \rightarrow 7 \rightarrow 9 \rightarrow 11$, $0 \rightarrow 3 \rightarrow 4 \rightarrow 7 \rightarrow 9 \rightarrow 11$, $0 \rightarrow 1 \rightarrow 3 \rightarrow 4 \rightarrow 7 \rightarrow 9 \rightarrow 11$, $0 \rightarrow 1 \rightarrow 3 \rightarrow 4 \rightarrow 7 \rightarrow 8 \rightarrow 9 \rightarrow 11$ and $0 \rightarrow 1 \rightarrow 3 \rightarrow 4 \rightarrow 7 \rightarrow 8 \rightarrow 9 \rightarrow 10 \rightarrow 11$. This is a significantly richer set of scale features than what OSNet can express. The search process also reveals that the architectural differences are tied to the positional roles of each block within the network. A comparison between "conv4_2" and "conv2_1" shows that since "conv2_1" is closer to the input, it is designed with shorter paths to efficiently extract low-level features. In contrast, "conv4_2" requires a longer path to effectively capture high-level semantic features.

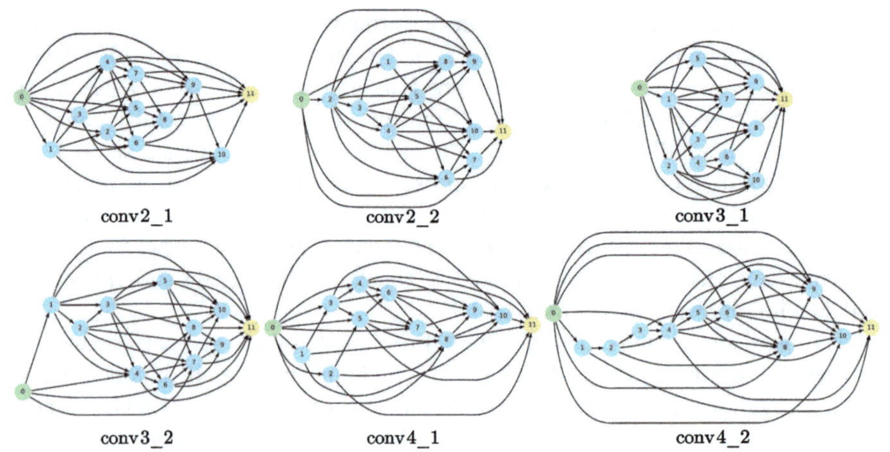

Fig. 3.7 The best searched architecture

3.4.4 Comparison with State-of-the-Art Methods

We compare our model with other state-of-the-art methods on three datasets, with the results summarized in Table 3.5. Without ImageNet pre-training, our model achieves competitive performance across all benchmarks: 94.5% Rank-1 and 85.3% mAP on Market1501; 63.3% Rank-1 and 60.5% mAP on CUHK03 (labeled); 58.6% Rank-1 and 58.0% mAP on CUHK03 (detected); and 74.3% Rank-1 and 48.6% mAP on MSMT17. Notably, our model maintains the same compact size as OSNet, with the fewest parameters among all compared methods, while significantly outperforming OSNet. It even achieves competitive results when compared to some models that have been pre-trained on ImageNet.

When pre-trained on the ImageNet dataset, our model's performance improves to 95.4% Rank-1 and 88.7% mAP on Market1501; 80.0% Rank-1 and 77.3% mAP on CUHK03 (labeled); 78.2% Rank-1 and 75.0% mAP on CUHK03 (detected); and 81.3% Rank-1 and 60.5% mAP on MSMT17. This pre-trained model achieves state-of-the-art performance on all three datasets, demonstrating a substantial improvement over the non-pre-trained version. It is important to highlight that most competing models are based on larger backbones like ResNet-50 or GoogleNet, and thus have significantly more parameters. Similar to OSNet, our model contains only 2.2M parameters, yet it consistently delivers superior performance. Our approach, like Auto-ReID, uses AutoML for architecture search. However, the model obtained by Auto-ReID has 13.1M parameters, nearly six times the size of ours. Our model consistently outperforms Auto-ReID, both when trained from scratch and when pre-trained on ImageNet, which suggests that our search space is more effective. Furthermore, our model is orthogonal to methods like MGN (Wang et al. 2018a), which uses

Table 3.5 Comparison to state-of-the-art methods across three datasets. Performance achieved by our method is highlighted in red, while the best performance reported by other existing methods is shown in blue

Method	Params(M)	Market1501		CUHK03				MSMT17	
				Labeled		Detected			
		R1	mAP	R1	mAP	R1	mAP	R1	mAP
Trained from scratch									
BraidNet (Wang et al. 2018c)	–	83.7	69.5	–	–	–	–	–	–
HAN (Li et al. 2018)	4.5	91.2	75.7	44.4	41.0	41.7	38.6	–	–
OSNet (Zhou et al. 2019)	2.2	93.6	81.0	–	–	57.1	54.2	71.0	43.3
Auto-ReID (Quan et al. 2019)	13.1	90.7	74.6	–	–	–	–	–	–
ASNet(ours)	2.2	94.5	85.3	63.3	60.5	58.6	58.0	74.3	48.6
Pre-trained on ImageNet									
SVDNet (Sun et al. 2017)	>23.5	82.3	62.1	–	–	41.5	37.3	–	–
PDC (Su et al. 2017)	>6.8	84.1	63.4	–	–	–	–	58.0	29.7
HAP2S (Yu et al. 2018)	>23.5	84.6	69.4	–	–	–	–	–	–
DPFL (Chen et al. 2017)	>6.8	88.6	72.6	43.0	40.5	40.7	37.0	–	–
DaRe (Wang et al. 2018d)	>23.5	89.0	76.0	66.1	61.6	63.3	59.0	–	–
PNGAN (Qian et al. 2018)	>23.5	89.4	72.6	–	–	–	–	–	–
GLAD (Wei et al. 2017)	>6.8	89.9	73.9	–	–	–	–	61.4	34.0
KPM (Shen et al. 2018a)	>23.5	90.1	75.3	–	–	–	–	–	–
MLFN (Chang et al. 2018)	32.5	90.0	74.3	54.7	49.2	52.8	47.8	–	–
FDGAN (Ge et al. 2018)	>23.5	90.5	77.7	–	–	–	–	–	–
DuATM (Si et al. 2018)	>7.0	91.4	76.6	–	–	–	–	–	–
Bilinear (Suh et al. 2018)	>6.8	91.7	79.6	–	–	–	–	–	–
G2G (Shen et al. 2018b)	>23.5	92.7	82.5	–	–	–	–	–	–
DeepCRF (Chen et al. 2018)	26.1	93.5	81.6	–	–	–	–	–	–
PCB (Sun et al. 2018)	27.2	93.8	81.6	–	–	63.7	57.5	68.2	40.4
SGGNN (Shen et al. 2018c)	>23.5	92.3	82.8	–	–	–	–	–	–
Mancs (Wang et al. 2018e)	>25.1	93.1	82.3	69.0	63.9	65.5	60.5	–	–
AANet (Tay et al. 2019)	>23.5	93.9	83.4	–	–	–	–	–	–
CAMA (Yang et al. 2019)	>23.5	94.7	84.5	70.1	66.5	66.6	64.2	–	–
IANet (Hou et al. 2019)	>23.5	94.4	83.1	–	–	–	–	75.5	46.8
DGNet (Zheng et al. 2019a)	>23.5	94.8	86.0	–	–	65.6	61.1	77.2	52.3
MGN (Wang et al. 2018a)	70.4	95.7	86.9	68.0	67.4	66.8	66.0	–	–
BagOfTricks (Luo et al. 2019)	25.6	94.5	85.9	–	–	–	–	–	–
OSNet (Zhou et al. 2019)	2.2	94.8	86.7	–	–	72.3	67.8	79.1	55.1
MHN (Chen et al. 2019b)	29.8	95.1	85.0	77.2	72.4	71.1	65.4	–	–
ABD-Net (Chen et al. 2019a)	>23.5	95.6	88.3	–	–	–	–	82.3	60.8
Auto-ReID (Quan et al. 2019)	13.1	94.5	85.1	77.9	73.0	73.3	69.3	78.2	52.5
ASNet(ours)	2.2	95.4	88.7	80.0	77.3	78.2	75.0	82.0	61.4

Table 3.6 Comparison with state-of-the-art methods on the Market1501 dataset, incorporating the re-ranking technique from (Zhong et al. 2017)

Method	Backbone	Params(M)	Market1501	
			R1	mAP
TriNet (Hermans et al. 2017)	ResNet-50	25.1	86.7	81.1
AOS (Huang et al. 2018)	ResNet-50	>25.1	88.7	83.3
AACN (Xu et al. 2018)	GoogleNet	>8.0	88.7	83.0
PSE+ECN (Sarfraz et al. 2018)	ResNet-50	>25.1	90.3	84.0
PCB (Sun et al. 2018)	ResNet-50	27.2	95.1	91.9
Auto-ReID (Quan et al. 2019)	Auto	13.1	95.4	94.2
ASNet(Ours)	Auto	**2.2**	**96.1**	**94.1**

a multi-branch architecture. A similar multi-branch design could be created based on our ASNet for further performance gains. Similarly, the orthogonality constraint from ABD-Net (Chen et al. 2019a) could also be integrated into our framework to further improve performance.

We also evaluated our model on the Market1501 dataset using the re-ranking method (Zhong et al. 2017), with results in Table 3.6. Despite having only 2.2M parameters, our model achieves a competitive mAP with Auto-ReID, while its Rank-1 accuracy reaches 96.1%, which is 0.7% higher than that of Auto-ReID.

3.5 Conclusion

In this chapter, we presented a novel method for automatically searching for multi-scale representations for person re-identification. The core of our approach is the Auto-Scale Block (ASB), which is composed of Lite 3×3 and RCB 1×1 operations. These ASBs are then used to form our macro network architecture, the Auto-Scale Network (ASNet), which consists of six such blocks. To address key challenges in the architecture search process, we introduced three types of regularization: entropy regularization to alleviate the discretization gap, validity regularization to ensure a valid path, and consistent regularization to eliminate meaningless edges. Finally, we demonstrated the effectiveness of our searched model on three common datasets. The results show that our model achieves state-of-the-art performance while maintaining the compact size of OSNet with only 2.2M parameters.

References

Alexander Hermans, Lucas Beyer, and Bastian Leibe. In defense of the triplet loss for person re-identification. *arXiv preprint* arXiv:1703.07737, 2017.

Andrew Brock, Theodore Lim, James M. Ritchie, and Nick Weston. SMASH: one-shot model architecture search through hypernetworks. In *6th International Conference on Learning Representations, ICLR 2018, Vancouver, BC, Canada, April 30 - May 3, 2018, Conference Track Proceedings*. OpenReview.net, 2018.

Barret Zoph and Quoc V. Le. Neural architecture search with reinforcement learning. In *5th International Conference on Learning Representations, ICLR 2017, Toulon, France, April 24-26, 2017, Conference Track Proceedings*. OpenReview.net, 2017.

Barret Zoph, Vijay Vasudevan, Jonathon Shlens, and Quoc V Le. Learning transferable architectures for scalable image recognition. In *Proceedings of the IEEE conference on computer vision and pattern recognition*, pages 8697–8710, 2018.

Binghui Chen, Weihong Deng, and Jiani Hu. Mixed high-order attention network for person re-identification. In *Proceedings of the IEEE International Conference on Computer Vision*, pages 371–381, 2019.

Cheng Wang, Qian Zhang, Chang Huang, Wenyu Liu, and Xinggang Wang. Mancs: A multi-task attentional network with curriculum sampling for person re-identification. In *Proceedings of the European Conference on Computer Vision (ECCV)*, pages 365–381, 2018.

Chi Su, Jianing Li, Shiliang Zhang, Junliang Xing, Wen Gao, and Qi Tian. Pose-driven deep convolutional model for person re-identification. In *Proceedings of the IEEE international conference on computer vision*, pages 3960–3969, 2017.

Chiat-Pin Tay, Sharmili Roy, and Kim-Hui Yap. Aanet: Attribute attention network for person re-identifications. In *Proceedings of the IEEE Conference on Computer Vision and Pattern Recognition*, pages 7134–7143, 2019.

Christian Szegedy, Vincent Vanhoucke, Sergey Ioffe, Jon Shlens, and Zbigniew Wojna. Rethinking the inception architecture for computer vision. In *Proceedings of the IEEE conference on computer vision and pattern recognition*, pages 2818–2826, 2016.

Dapeng Chen, Dan Xu, Hongsheng Li, Nicu Sebe, and Xiaogang Wang. Group consistent similarity learning via deep crf for person re-identification. In *Proceedings of the IEEE conference on computer vision and pattern recognition*, pages 8649–8658, 2018.

Feng Zheng, Cheng Deng, Xing Sun, Xinyang Jiang, Xiaowei Guo, Zongqiao Yu, Feiyue Huang, and Rongrong Ji. Pyramidal person re-identification via multi-loss dynamic training. In *Proceedings of the IEEE Conference on Computer Vision and Pattern Recognition*, pages 8514–8522, 2019.

Gabriel Bender, Pieter-Jan Kindermans, Barret Zoph, Vijay Vasudevan, and Quoc V. Le. Understanding and simplifying one-shot architecture search. In Jennifer G. Dy and Andreas Krause, editors, *Proceedings of the 35th International Conference on Machine Learning, ICML 2018, Stockholmsmässan, Stockholm, Sweden, July 10-15, 2018*, volume 80 of *Proceedings of Machine Learning Research*, pages 549–558. PMLR, 2018.

Guanshuo Wang, Yufeng Yuan, Xiong Chen, Jiwei Li, and Xi Zhou. Learning discriminative features with multiple granularities for person re-identification. In *Proceedings of the 26th ACM international conference on Multimedia*, pages 274–282, 2018.

Guanshuo Wang, Yufeng Yuan, Xiong Chen, Jiwei Li, and Xi Zhou. Learning discriminative features with multiple granularities for person re-identification. In Susanne Boll, Kyoung Mu Lee, Jiebo Luo, Wenwu Zhu, Hyeran Byun, Chang Wen Chen, Rainer Lienhart, and Tao Mei, editors, *2018 ACM Multimedia Conference on Multimedia Conference, MM 2018, Seoul, Republic of Korea, October 22-26, 2018*, pages 274–282. ACM, 2018.

Hanxiao Liu, Karen Simonyan, and Yiming Yang. DARTS: differentiable architecture search. In *7th International Conference on Learning Representations, ICLR 2019, New Orleans, LA, USA, May 6-9, 2019*. OpenReview.net, 2019.

Hao Luo, Youzhi Gu, Xingyu Liao, Shenqi Lai, and Wei Jiang. Bag of tricks and a strong baseline for deep person re-identification. In *Proceedings of the IEEE Conference on Computer Vision and Pattern Recognition Workshops*, pages 0–0, 2019.

Houjing Huang, Dangwei Li, Zhang Zhang, Xiaotang Chen, and Kaiqi Huang. Adversarially occluded samples for person re-identification. In *Proceedings of the IEEE Conference on Computer Vision and Pattern Recognition*, pages 5098–5107, 2018.

Jianlou Si, Honggang Zhang, Chun-Guang Li, Jason Kuen, Xiangfei Kong, Alex C Kot, and Gang Wang. Dual attention matching network for context-aware feature sequence based person re-identification. In *Proceedings of the IEEE Conference on Computer Vision and Pattern Recognition*, pages 5363–5372, 2018.

Jing Xu, Rui Zhao, Feng Zhu, Huaming Wang, and Wanli Ouyang. Attention-aware compositional network for person re-identification. In *Proceedings of the IEEE conference on computer vision and pattern recognition*, pages 2119–2128, 2018.

Kaiming He, Xiangyu Zhang, Shaoqing Ren, and Jian Sun. Deep residual learning for image recognition. In *Proceedings of the IEEE conference on computer vision and pattern recognition*, pages 770–778, 2016.

Kaiyang Zhou, Yongxin Yang, Andrea Cavallaro, and Tao Xiang. Omni-scale feature learning for person re-identification. In *Proceedings of the IEEE International Conference on Computer Vision*, pages 3702–3712, 2019.

Liang Zheng, Liyue Shen, Lu Tian, Shengjin Wang, Jingdong Wang, and Qi Tian. Scalable person re-identification: A benchmark. In *Proceedings of the IEEE international conference on computer vision*, pages 1116–1124, 2015.

Longhui Wei, Shiliang Zhang, Hantao Yao, Wen Gao, and Qi Tian. Glad: Global-local-alignment descriptor for pedestrian retrieval. In *Proceedings of the 25th ACM international conference on Multimedia*, pages 420–428, 2017.

Longhui Wei, Shiliang Zhang, Wen Gao, and Qi Tian. Person transfer gan to bridge domain gap for person re-identification. In *Proceedings of the IEEE conference on computer vision and pattern recognition*, pages 79–88, 2018.

M Saquib Sarfraz, Arne Schumann, Andreas Eberle, and Rainer Stiefelhagen. A pose-sensitive embedding for person re-identification with expanded cross neighborhood re-ranking. In *Proceedings of the IEEE Conference on Computer Vision and Pattern Recognition*, pages 420–429, 2018.

Rui Yu, Zhiyong Dou, Song Bai, Zhaoxiang Zhang, Yongchao Xu, and Xiang Bai. Hard-aware point-to-set deep metric for person re-identification. In *Proceedings of the European Conference on Computer Vision (ECCV)*, pages 188–204, 2018.

Ruibing Hou, Bingpeng Ma, Hong Chang, Xinqian Gu, Shiguang Shan, and Xilin Chen. Interaction-and-aggregation network for person re-identification. In *Proceedings of the IEEE Conference on Computer Vision and Pattern Recognition*, pages 9317–9326, 2019.

Ruijie Quan, Xuanyi Dong, Yu Wu, Linchao Zhu, and Yi Yang. Auto-reid: Searching for a part-aware convnet for person re-identification. In *2019 IEEE/CVF International Conference on Computer Vision, ICCV 2019, Seoul, Korea (South), October 27 - November 2, 2019*, pages 3749–3758. IEEE, 2019.

Tianlong Chen, Shaojin Ding, Jingyi Xie, Ye Yuan, Wuyang Chen, Yang Yang, Zhou Ren, and Zhangyang Wang. Abd-net: Attentive but diverse person re-identification. In *Proceedings of the IEEE International Conference on Computer Vision*, pages 8351–8361, 2019.

Wei Li, Rui Zhao, Tong Xiao, and Xiaogang Wang. Deepreid: Deep filter pairing neural network for person re-identification. In *Proceedings of the IEEE conference on computer vision and pattern recognition*, pages 152–159, 2014.

Wei Li, Xiatian Zhu, and Shaogang Gong. Harmonious attention network for person re-identification. In *Proceedings of the IEEE conference on computer vision and pattern recognition*, pages 2285–2294, 2018.

Wenjie Luo, Yujia Li, Raquel Urtasun, and Richard Zemel. Understanding the effective receptive field in deep convolutional neural networks. In *Advances in neural information processing systems*, pages 4898–4906, 2016.

Wenjie Yang, Houjing Huang, Zhang Zhang, Xiaotang Chen, Kaiqi Huang, and Shu Zhang. Towards rich feature discovery with class activation maps augmentation for person re-identification. In *Proceedings of the IEEE Conference on Computer Vision and Pattern Recognition*, pages 1389–1398, 2019.

Xiaobin Chang, Timothy M Hospedales, and Tao Xiang. Multi-level factorisation net for person re-identification. In *Proceedings of the IEEE Conference on Computer Vision and Pattern Recognition*, pages 2109–2118, 2018.

Xuelin Qian, Yanwei Fu, Tao Xiang, Wenxuan Wang, Jie Qiu, Yang Wu, Yu-Gang Jiang, and Xiangyang Xue. Pose-normalized image generation for person re-identification. In *Proceedings of the European conference on computer vision (ECCV)*, pages 650–667, 2018.

Yan Wang, Lequn Wang, Yurong You, Xu Zou, Vincent Chen, Serena Li, Gao Huang, Bharath Hariharan, and Kilian Q Weinberger. Resource aware person re-identification across multiple resolutions. In *Proceedings of the IEEE Conference on Computer Vision and Pattern Recognition*, pages 8042–8051, 2018.

Yanbei Chen, Xiatian Zhu, and Shaogang Gong. Person re-identification by deep learning multi-scale representations. In *Proceedings of the IEEE International Conference on Computer Vision Workshops*, pages 2590–2600, 2017.

Yang Fu, Yunchao Wei, Yuqian Zhou, Honghui Shi, Gao Huang, Xinchao Wang, Zhiqiang Yao, and Thomas Huang. Horizontal pyramid matching for person re-identification. In *Proceedings of the AAAI Conference on Artificial Intelligence*, volume 33, pages 8295–8302, 2019.

Yantao Shen, Hongsheng Li, Shuai Yi, Dapeng Chen, and Xiaogang Wang. Person re-identification with deep similarity-guided graph neural network. In *Proceedings of the European conference on computer vision (ECCV)*, pages 486–504, 2018.

Yantao Shen, Hongsheng Li, Tong Xiao, Shuai Yi, Dapeng Chen, and Xiaogang Wang. Deep group-shuffling random walk for person re-identification. In *Proceedings of the IEEE conference on computer vision and pattern recognition*, pages 2265–2274, 2018.

Yantao Shen, Tong Xiao, Hongsheng Li, Shuai Yi, and Xiaogang Wang. End-to-end deep kronecker-product matching for person re-identification. In *Proceedings of the IEEE Conference on Computer Vision and Pattern Recognition*, pages 6886–6895, 2018.

Yicheng Wang, Zhenzhong Chen, Feng Wu, and Gang Wang. Person re-identification with cascaded pairwise convolutions. In *Proceedings of the IEEE Conference on Computer Vision and Pattern Recognition*, pages 1470–1478, 2018.

Yifan Sun, Liang Zheng, Weijian Deng, and Shengjin Wang. Svdnet for pedestrian retrieval. In *Proceedings of the IEEE International Conference on Computer Vision*, pages 3800–3808, 2017.

Yifan Sun, Liang Zheng, Yi Yang, Qi Tian, and Shengjin Wang. Beyond part models: Person retrieval with refined part pooling (and a strong convolutional baseline). In *Proceedings of the European Conference on Computer Vision (ECCV)*, pages 480–496, 2018.

Yixiao Ge, Zhuowan Li, Haiyu Zhao, Guojun Yin, Shuai Yi, Xiaogang Wang, et al. Fd-gan: Pose-guided feature distilling gan for robust person re-identification. In *Advances in neural information processing systems*, pages 1222–1233, 2018.

Yumin Suh, Jingdong Wang, Siyu Tang, Tao Mei, and Kyoung Mu Lee. Part-aligned bilinear representations for person re-identification. In *Proceedings of the European Conference on Computer Vision (ECCV)*, pages 402–419, 2018.

Zhedong Zheng, Xiaodong Yang, Zhiding Yu, Liang Zheng, Yi Yang, and Jan Kautz. Joint discriminative and generative learning for person re-identification. In *Proceedings of the IEEE Conference on Computer Vision and Pattern Recognition*, pages 2138–2147, 2019.

Zhun Zhong, Liang Zheng, Donglin Cao, and Shaozi Li. Re-ranking person re-identification with k-reciprocal encoding. In *Proceedings of the IEEE Conference on Computer Vision and Pattern Recognition*, pages 1318–1327, 2017.

Chapter 4
Multi-Scale Multi-Branch Architecture Search for Person Re-identification

4.1 Introduction

Person re-identification (ReID) is a prominent computer vision task that addresses the challenge of retrieving images of a specific person across a network of non-overlapping cameras. This problem is inherently difficult due to significant variations in pose, viewpoint, lighting, and occlusions, compounded by the presence of both intra-class variations and high inter-class similarity between different individuals. Consequently, ReID has become a highly active and challenging area of research.

With the rapid development of deep learning, deep convolutional neural networks (DCNNs) have achieved remarkable success in various computer vision tasks due to their powerful feature representation capabilities. In the field of ReID, a growing number of methods leverage DCNNs to extract robust features, leading to breakthrough performance with high identification rates (Chen et al. 2018; Chang et al. 2018; Li et al. 2018; Sun et al. 2018; Wang et al. 2018a). However, as noted in previous analyses (Quan et al. 2019), these methods are predominantly based on backbones such as ResNet, Inception, and VGG. These architectures were originally designed for classification problems, making them less than ideal for metric learning tasks like ReID, where the goal is to learn a discriminative distance metric rather than a simple categorical output.

To address this limitation, Neural Architecture Search (NAS) has emerged as a method to automatically discover network backbones tailored to specific tasks. In this context, Auto-ReID (Quan et al. 2019) pioneered the use of NAS to search for a ReID network architecture. It is noteworthy that Auto-ReID primarily searches for an architecture optimized for extracting global features, rather than local, part-level features. In contrast, part-based models (Sun et al. 2018) have demonstrated significant efficacy by uniformly partitioning the backbone's feature maps along the vertical direction to extract local information from each body part, leveraging human body structure prior knowledge. The simplicity and effectiveness of this approach have made it a popular baseline. Building on this, MGN (Wang et al. 2018a) expanded

H. Gu et al., *Automated Machine Learning for Person Re-Identification*, Intelligent Perception and Information Processing,
https://doi.org/10.1007/978-981-95-3433-3_4

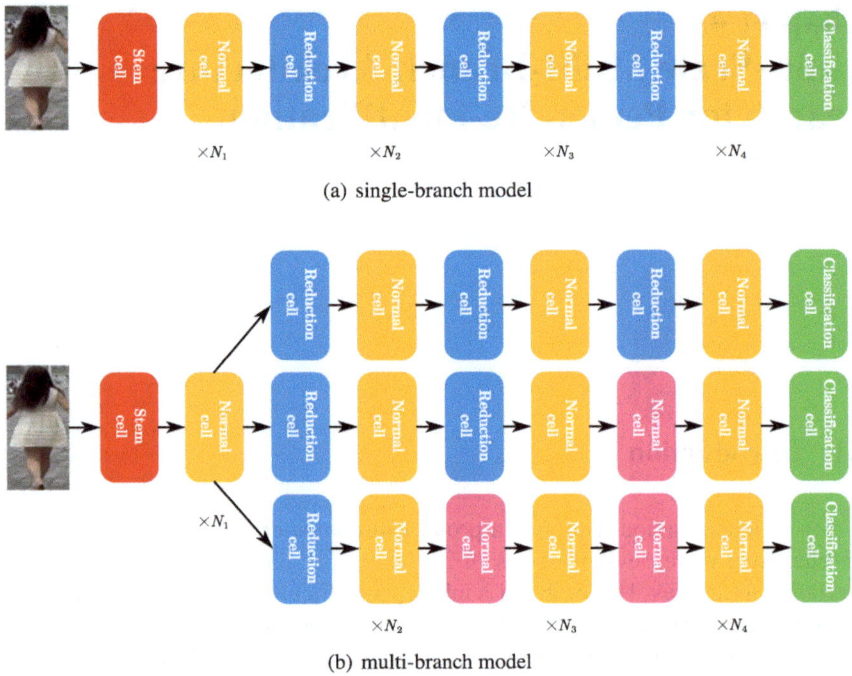

Fig. 4.1 Different macro network architecture

the network into a multi-branch structure, resulting in a substantial leap in person re-identification accuracy. Given the inherent advantages of multi-branch and part-based models in ReID tasks, and the limitations of manually designing them based on classification-oriented backbones, it is imperative to develop a method that can automatically generate a multi-branch network architecture specifically for the ReID task.

Within the search framework of Auto-ReID, this chapter introduces a novel method named Auto-ReID+, aiming to broaden the application of Neural Architecture Search (NAS) for the person re-identification (ReID) task. The primary contribution of Auto-ReID+ is the design of a new macro network architecture for ReID search (Fig. 4.1b). Specifically, this macro-architecture is composed of three distinct branches, each producing feature maps of different spatial dimensions. Leveraging the effective receptive field theory (Luo et al. 2016), which states that a smaller feature map scale corresponds to a larger effective receptive field per neuron, this design naturally facilitates the extraction of multi-scale features. Furthermore, we uniformly partition each feature map into a varying number of parts based on its size, allowing us to capture local features at diverse scales. This is complemented by the preservation of a global feature representation across all scales. In addition to these architectural innovations, we augment the Auto-ReID search strategy with an early stopping mechanism. This addition not only significantly accelerates the search process but also enhances the overall performance of the resulting network. In summary, our key contributions are as follows:

(1) To the best of our knowledge, this is the first work to apply neural architecture search to discover a multi-branch network architecture for the person re-identification task.
(2) We propose a novel multi-branch macro-architecture that synergistically integrates multi-scale, local, and global features, effectively capturing the strengths of existing person re-identification models.
(3) The proposed early stopping mechanism substantially enhances both the efficiency of the search process and the ultimate effectiveness of the discovered model.
(4) Extensive experiments on three person re-identification benchmarks demonstrate that our model achieves highly competitive results even without ImageNet pre-training. When pre-trained, our approach achieves **state-of-the-art** performance with a parameter count of just 5.7 million.

4.2 Related Works

4.2.1 Person Re-identification

Deep learning techniques have driven significant advancements in the field of person re-identification (ReID) (Sun et al. 2017, 2018; Wang et al. 2018a, b; Su et al. 2017; Yu et al. 2018; Chen et al. 2017; Liu et al. 2016, 2019a, b). These models can be broadly categorized based on their feature extraction approach: models that leverage global features and those that rely on local features. Global feature-based models directly use a single feature vector extracted from the entire image by a backbone network (such as ResNet50 and InceptionV3). While straightforward, this approach often yields features that are highly sensitive to misalignments and occlusions. This can lead to overfitting on smaller datasets and suboptimal overall performance, often requiring extensive "bag of tricks" techniques (Luo et al. 2019) to achieve competitive results.

In contrast, models based on local features are capable of learning more abundant and discriminative representations. A seminal work in this category, PCB (Sun et al. 2018) by Sun et al., leverages the prior knowledge of human body structure to roughly partition the backbone network's feature map uniformly and vertically, thereby obtaining part-specific local features. This simple yet effective design has been widely adopted in a large number of subsequent research efforts (Li et al. 2018; Wang et al. 2018a; Zheng et al. 2019).

From a network architecture perspective, ReID models can also be classified as either single-branch or multi-branch designs (Li et al. 2018; Wang et al. 2018a; Cheng et al. 2016). The aforementioned PCB model and nearly all global feature-based models fall into the single-branch category. Among multi-branch models, MGN

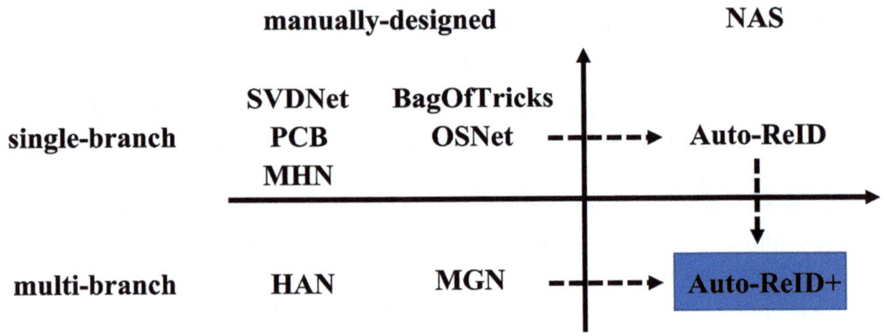

Fig. 4.2 This illustration presents a classification framework for Re-ID, categorized by the number of network branches and the design methodology

(Wang et al. 2018a) is a typical example, which introduces multi-scale information by incorporating a multi-branch structure. This model architecture once again set a new record for the ReID task, demonstrating that multi-branch models possess a significant innate advantage in the ReID domain.

All the models mentioned above are a result of manual design, and their backbones are often adapted from architectures originally intended for classification. As highlighted by the Auto-ReID framework, there is a fundamental difference between a classification task and the metric learning inherent in ReID, underscoring the need for a dedicated, task-specific backbone architecture. The Auto-ReID method is specifically dedicated to searching for a single-branch model for extracting global features. As previously discussed, a multi-branch structure offers significant advantages in ReID tasks. Thus, the Auto-ReID+ method proposed in this chapter is dedicated to searching for a multi-branch model capable of extracting both global and local features. Our placement within the existing research landscape is illustrated in Fig. 4.2.

4.2.2 Neural Architecture Search

Neural Architecture Search (NAS) is a prominent and rapidly evolving subfield of automated machine learning (AutoML), with the primary objective of automating the discovery of high-performance network architectures. This automation is designed to minimize the need for manual design and extensive domain expertise, thereby democratizing the development of deep neural networks. The process of designing an optimal network architecture is often a complex and time-consuming task, involving significant human effort and trial-and-error. NAS addresses this by transforming the design problem into an optimization problem, where the goal is to find the best-performing architecture given a specific task and dataset. The early explorations into NAS, as described in this chapter, were often computationally prohibitive. These initial methods typically searched on a small proxy dataset and then transferred the discovered architecture to a larger target dataset (Liu et al. 2019c; Zoph and Le 2017; Zoph et al. 2018). For instance, Zoph et al. (2017, 2018) pioneered the use of reinforcement learning (RL) to select architectural components from a predefined search space. A controller network would predict a sequence of tokens representing an architecture, which would then be instantiated, trained from scratch, and evaluated. The performance of the discovered architecture was used as a reward signal to update the controller. However, the search efficiency of such methods was extremely low, often requiring thousands of GPU days to find a suitable architecture. The fundamental challenge lay in the combinatorial nature of the search space, where each candidate architecture had to be fully trained and validated independently, making the process unscalable for modern, large-scale deep learning tasks.

To address these computational challenges and make NAS more accessible, a new generation of more efficient approaches emerged. A significant shift came with the exploration of one-shot NAS methods, pioneered by works such as those from Brock et al. (2018) and Bender et al. (2018). Instead of training and evaluating each architecture individually, these methods propose training a single, large "supernet" that contains all possible architectures within the search space as sub-graphs. The search problem is then reduced to finding the optimal sub-network within this shared supernet, a process that is orders of magnitude faster.

A major breakthrough, however, was the introduction of differentiable NAS (DNAS), most famously exemplified by DARTS (Differentiable Architecture Search) by Liu et al. (2019c). This groundbreaking work introduced the concept of relaxing the discrete architectural search space to a continuous one. Instead of making a hard choice between operations like convolution and pooling at each node, DARTS represents the choice as a weighted combination of all possible operations. The architectural weights (which determine the mixing ratios) become learnable parameters, allowing the search to be conducted within the same differentiable framework (such as TensorFlow or PyTorch) used for network training. This transformation allows the search for an optimal architecture to be performed using gradient descent, just like the network's weights themselves. This breakthrough greatly enhances search efficiency; a search can often be completed in just a few GPU days, a stark contrast to the thousands previously required. This efficiency makes it feasible to abandon the proxy paradigm and directly discover a robust CNN architecture on a target dataset. This is particularly beneficial for complex and specific domains like person re-identification, where the discovered architectures can be tailored to address the unique challenges of the task.

4.3 Methodology

In this section, we present a detailed methodology for discovering a high-performance, multi-branch, multi-scale Convolutional Neural Network (CNN) architecture for the person Re-Identification (ReID) task. The process is broken down into three key steps. First, we establish the foundational principles of Neural Architecture Search (NAS) in Sect. 4.3.1. Subsequently, we detail the design of our proposed multi-scale and multi-branch macro-architecture in Sect. 4.3.2. Finally, we introduce the integration of an early stopping mechanism, which significantly enhances the efficiency and effectiveness of the search procedure, as outlined in Sect. 4.3.3.

4.3.1 Preliminary

Neural Architecture Search (NAS) methodologies are broadly categorized into two primary types of search spaces (Zoph and Le 2017; Zoph et al. 2018; Liu et al. 2018; Chu et al. 2019; Xu et al. 2020; Liang et al. 2019a): macro search space and micro search space. The former involves designing the complete network architecture directly through a generative process, yielding an end-to-end model. In contrast, the latter focuses on discovering reusable architectural motifs, or "cells" which are then stacked repeatedly to form the entire network. The micro search space approach is generally favored due to its computational tractability and often yields superior performance. In this chapter, our methodology similarly adopts the micro search space paradigm, where the comprehensive network structure is constructed by the iterative stacking of multiple learned cells.

Following the precedent set by DARTS (Liu et al. 2019c), we construct the ultimate network architecture utilizing distinct types of cells: normal cells and reduction cells. Each cell is conceptualized as a directed acyclic graph (DAG) comprising N ordered nodes. Each node within this graph corresponds to an intermediate feature map. An edge, denoted as $e^{(i,j)}$, originating from node i and terminating at node j, applies an operation to the input feature $x^{(i)}$, with its resulting output being $o^{(i,j)}(x^{(i)})$. An intermediate node j aggregates the outputs from all preceding nodes connected by incoming edges, as expressed by the following summation:

$$x^{(j)} = \sum_{i<j} o^{(i,j)}(x^{(i)}). \tag{4.1}$$

Let $O = \{o_1^{(i,j)}, o_2^{(i,j)}, ..., o_M^{(i,j)}\}$ represent the collection of M potential operations (e.g., convolution, max-pooling, zero operation) applicable to an edge $e^{(i,j)}$. DARTS transforms this discrete selection into a continuous, differentiable form by applying a softmax function over all operations in O to produce a weighted combination of outputs:

$$\overline{o}^{(i,j)}(x) = \sum_{o \in O} \frac{\exp(\alpha_o^{(i,j)})}{\sum_{o' \in O} \exp(\alpha_{o'}^{(i,j)})} o(x), \tag{4.2}$$

where the operation mixing weights for a specific pair of nodes (i, j) are governed by a vector $\alpha^{(i,j)}$ of dimension $|O|$. Consequently, the entire architectural parameter set is defined as $\alpha = \{\alpha^{(i,j)}\}$. The objective of the architecture search then becomes identifying the optimal α that minimizes the validation loss $\mathcal{L}_{val}(w^*(\alpha), \alpha)$. In this context, $w^*(\alpha)$ denotes the optimized network weights w that minimize the training loss $\mathcal{L}_{train}(w, \alpha)$ under the current architectural configuration α. This implies that during the optimization of the architectural parameters α, the network weights w must be optimized to convergence. In summary, this constitutes a bi-level optimization problem:

$$\begin{aligned} \alpha^* &= \arg\min_{\alpha} \mathcal{L}_{val}\left(w^*\left(\alpha\right), \alpha\right) \\ \text{s.t.} w^*\left(\alpha\right) &= \arg\min_{w} \mathcal{L}_{train}\left(w, \alpha\right) \end{aligned}. \tag{4.3}$$

However, directly optimizing $\mathcal{L}_{val}(w^*(\alpha), \alpha)$ presents a significant challenge. DARTS (Liu et al. 2019c) addresses this challenge through an alternating optimization scheme. Specifically, prior to each update of the architectural parameters α using the validation dataset, the network weights w are first refined based on the training dataset. By alternately updating α and w in this manner, the aforementioned bi-level optimization problem is iteratively approximated. Further details on this process can be found in steps 4 to 7 of Algorithm 4.1.

The final step in the DARTS-based search process is to derive a discrete, concrete architecture from the learned continuous parameters. After the architectural weights α have been optimized, the continuous relaxation of the cell needs to be resolved into a fixed set of operations. To achieve this, for each intermediate node in the cell, we can retain the top-k strongest operations. These are selected from all non-zero candidate operations whose edges originate from distinct previous nodes. In this chapter, we set $k = 2$, following the methodology proposed by Liu et al. (2019c).

The strength of an operation, which determines its ranking, is defined by the magnitude of its architectural weight. Specifically, it is given by the normalized exponential of the weight:

$$\frac{\exp(\alpha_o^{(i,j)})}{\sum_{o' \in O} \exp(\alpha_{o'}^{(i,j)})}. \tag{4.4}$$

This metric reflects the contribution of operation o to the mixed output on edge $e^{(i,j)}$. Consistent with the original DARTS implementation (Liu et al. 2019c), zero operations are excluded from consideration during this selection process, as they represent a null connection and are not part of the final architecture.

4.3.2 Macro Network Architecture

Prior to initiating the search for optimal cells, it is essential to pre-define the overarching macro network architecture. A common practice in existing literature, as demonstrated by early work (Quan et al. 2019), is to employ a general macro structure, often inspired by classic architectures like ResNet or VGG, as illustrated in Fig. 4.1a. Within this predefined framework, certain components are fixed beforehand. For instance, a stem cell is typically set at the network's beginning, often consisting of an initial 3 × 3 convolutional layer. The architectural search parameters, denoted as $(\alpha_{normal}, \alpha_{reduction})$, are designed with weight-sharing in mind: α_{normal} is shared across all normal cells, while $\alpha_{reduction}$ is shared by all reduction cells. Furthermore, the classification cell can also be tailored and preset according to the specific requirements of different tasks (Quan et al. 2019; Ghiasi et al. 2019; Chen et al. 2019a).

Effective feature extraction, particularly for tasks like person re-identification, often requires capturing visual information across multiple granularities. To achieve this, a multi-branch and multi-scale network topology proves to be highly effective (Wang et al. 2018a). Consequently, we propose a novel multi-scale and multi-branch macro network architecture, as depicted in Fig. 4.1b. The core principle of this design is that different branches are configured to maintain distinct resolutions, thereby guaranteeing the extraction of features at varying scales. This network is composed of four unique cell types: a stem cell, normal cells, reduction cells, and a final classification cell.

4.3.2.1 Stem Cell

The stem cell, positioned at the beginning of the network, is a crucial component responsible for initial feature extraction and downsampling. Following a common practice in many state-of-the-art architectures like ResNet, our design for the stem cell incorporates a 7 × 7 convolution with a stride of 2, followed by a 3 × 3 max-pooling layer, also with a stride of 2.

This specific configuration serves a critical purpose: it rapidly reduces the spatial dimensions of the input image. After passing through this stem cell, the spatial size is decreased to one-quarter of its original dimensions. This aggressive downsampling strategy is a deliberate design choice that ensures the computational complexity of the subsequent network layers remains manageable, thereby preventing the overall network's floating-point operations (FLOPs) from becoming prohibitively large.

4.3.2.2 Normal Cells and Reduction Cells

The architectures of the normal and reduction cells are determined by optimizing the architectural parameters, α_{normal} and $\alpha_{reduction}$, as defined in Eq. (4.3). As shown in Fig. 4.1b, the network begins with an initial block comprising N_1 normal cells. The feature map produced by this block is then used to extract multi-level, fine-grained features via three subsequent branches, each operating at a different resolution. Specifically, these three branches, from top to bottom, are designated as Branch 1, Branch 2, and Branch 3. The number of reduction cells (represented by the blue boxes in Fig. 4.1b) within each branch progressively decreases, from three in Branch 1 to one in Branch 3. In positions where a reduction cell is not present, a normal cell is used as a substitute (indicated by the pink boxes in Fig. 4.1b). Following either a reduction cell or a substitute normal cell, additional normal cells (also shown as blue boxes in Fig. 4.1b) are stacked.

It is important to emphasize a key distinction between these two types of normal cells. A reduction cell's primary function is to halve the spatial dimensions (height and width) of the feature map while simultaneously doubling the number of channels. Since the normal cells marked in pink replace reduction cells in certain positions, they are designed to perform a similar function of doubling the number of channels to maintain consistent feature dimensionality across branches. In contrast, the normal cells marked in blue do not alter the spatial dimensions or the number of channels. Despite their different functional roles, these two types of normal cells share the same architectural parameter, α_{normal}. This weight-sharing strategy is adopted to improve the overall search efficiency and effect, as supported by prior research (Yang et al. 2020).

4.3.2.3 Classification Cell

For the final classification cell, we adopt a structure inspired by the work of Wang et al. (2018a), as illustrated in Fig. 4.3. Given the varying resolutions of the feature maps from the different branches, our design incorporates a variable number of horizontal divisions; feature maps with higher resolution undergo more divisions, while those with lower resolution are subject to fewer. The specific configuration details for each branch are provided in Table 4.1.

The feature map obtained from Branch j is denoted as $f_0^{B_j}$. In order to extract both global and fine-grained information, we apply two distinct pooling strategies. For each branch j where $j \geq 2$, the feature map $f_0^{B_j}$ is partitioned into 2^{j-1} horizontal

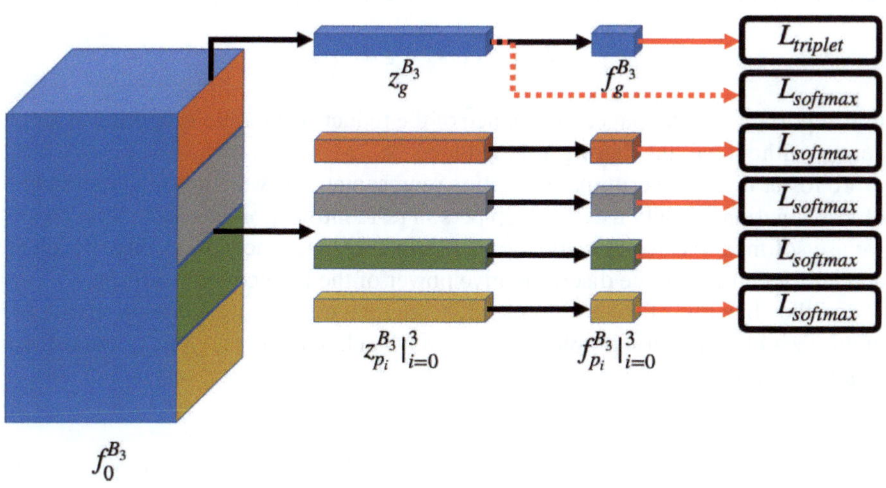

Fig. 4.3 Architecture of the classification cell (Branch 3)

Table 4.1 Comparison of configurations across the three branches. The input image size is fixed at 384 × 128. **Branch** specifies the branch identifier. **Part No.** indicates the number of partitions on the feature maps. **Map Size** denotes the output feature map dimensions for each branch. **Dims** represents the dimensionality and quantity of the output feature representations. **Feature** refers to the notation used for the output feature representations

Branch	Part No.	Map size	Dims	Feature
Branch 1	1	12 × 4	1 × 256	$f_g^{B_1}$
Branch 2	2	24 × 8	(1+2) × 256	$f_g^{B_2}, f_{p_i}^{B_2}\|_{i=0}^{1}$
Branch 3	4	48 × 16	(1+4) × 256	$f_g^{B_3}, f_{p_i}^{B_3}\|_{i=0}^{3}$

segments. A max-pooling (MAP) operation is then applied to each segment to yield the part-based features, $z_{p_i}^{B_j}$:

$$z_{p_i}^{B_j} = MAP(f_0^{B_j}[:, :, \delta i : \delta(i+1), :]),\ j \geq 2, i \in [0, ..., 2^{j-1} - 1], \tag{4.5}$$

where $\delta = \frac{H}{2^{j-1}}$ and H is the height of the feature map $f_0^{B_j}$.

Concurrently, for all three branches ($j \in [1, 2, 3]$), a global feature is obtained by applying a global max-pooling operation across the entire feature map of each branch:

$$z_g^{B_j} = MAP(f_0^{B_j}),\ j \in [1, 2, 3]. \tag{4.6}$$

To achieve a more compact and manageable feature representation, the set of all extracted features, $Z = \{z_g^{B_1}, z_g^{B_2}, z_{p_i}^{B_2}|_{i=0}^{1}, z_g^{B_3}, z_{p_i}^{B_3}|_{i=0}^{3}\}$, is projected to a lower-dimensional space. This projection is accomplished using a 1 × 1 convolutional layer followed by batch normalization, which reduces the feature dimension to 256. The resulting compact features are collected into a set $F = \{f_g^{B_1}, f_g^{B_2}, f_{p_i}^{B_2}|_{i=0}^{1}, f_g^{B_3}, f_{p_i}^{B_3}|_{i=0}^{3}\}$.

$$f = Reduce(z),\ f \in F, z \in Z. \tag{4.7}$$

It is important to note that each instance of the reduction module is applied independently, without weight sharing among them.

To foster highly discriminative feature representations, we employ a combination of loss functions widely used in deep person re-identification methods. Specifically, we use softmax loss for classification and triplet loss for metric learning, which are crucial for enhancing the discriminative power of the learned embeddings.

For the objective of basic classification, we treat the identity recognition task as a multi-class classification problem. For a given batch of features, $\{f_i\}|_{i=1}^{N}$, the softmax loss function is defined as

$$L_{softmax} = -\sum_{i=1}^{N} \log \frac{e^{w_{y_i}^T f_i}}{\sum_{c=1}^{C} e^{w_c{}^T f_i}}, \tag{4.8}$$

where w_c represents a weight vector corresponding to class c, y_i is the ground-truth label for the i-th sample, and C is the total number of classes. To ensure fundamental discriminative learning, we apply the softmax loss to a selection of the learned features, including the global features before the reduction module $\{z_g^{B_1}, z_g^{B_2}, z_g^{B_3}\}$ and the part features after their dimensionality reduction $\{f_{p_i}^{B_2}|_{i=0}^{1}, f_{p_i}^{B_3}|_{i=0}^{3}\}$.

To further enhance performance, all reduced global features, $\{f_g^{B_1}, f_g^{B_2}, f_g^{B_3}\}$, are trained using triplet loss to improve metric learning. We adopt a variant known as batch hard triplet loss, an advanced version of the original semi-hard triplet loss. This loss function is formulated as follows:

$$L_{triplet} = \sum_{i=1}^{P}\sum_{a=1}^{K}[m + \max_{p=1...K} \left\| f_a^{(i)} - f_p^{(i)} \right\|_2 - \min_{\substack{n=1...K \\ j=1...P \\ j \neq i}} \left\| f_a^{(i)} - f_n^{(j)} \right\|_2]_+, \tag{4.9}$$

where $f_a^{(i)}$, $f_p^{(i)}$, and $f_n^{(j)}$ represent features extracted from an anchor, a positive, and a negative sample, respectively. The margin hyperparameter, m, regulates the minimum separation required between intra-class and inter-class distances. A positive sample shares the same identity as the anchor, while a negative sample has a different identity. The triplet candidates are formed by identifying the hardest pairs: the furthest positive and closest negative samples within each mini-batch. Each mini-batch consists of P unique identities, with K images for each identity. This formulation of triplet loss enhances the robustness of the feature learning process, leading to improved performance.

The overall objective function for the network is a weighted combination of the classification and metric learning losses:

$$L = L_{softmax} + \lambda L_{triplet}, \tag{4.10}$$

where λ is a hyperparameter that balances the contribution of the two losses. It is crucial to note that this combined loss function is used for two purposes: first, to guide the update of both the network weights w and the architectural parameters α during the search process, and second, to train the final optimal network once its architecture has been determined.

The final macro-level structure of the network is summarized in Table 4.2, where C represents the number of channels in the first convolutional layer (conv1). The number of normal cells in blocks 0, 2, 4, and 6 is specified as N_1, N_2, N_3, and N_4, respectively.

Table 4.2 Overall macro-architecture of the proposed network, with an input image size of 384 × 128

Stage	Output			Network		
Conv1	192 × 64, C			7 × 7 conv, stride 2		
Pool	96 × 32, C			3 × 3 max-pooling, stride 2		
Block0	96 × 32, C			Normal cell $\times N_1$		
	Branch 1	Branch 2	Branch 3	Branch 1	Branch 2	Branch 3
Block1	48 × 16,2C	48 × 16,2C	48 × 16,2C	Reduction cell	Reduction cell	Reduction cell
Block2	48 × 16,2C	48 × 16,2C	48 × 16,2C	Normal cell $\times N_2$	Normal cell $\times N_2$	Normal cell $\times N_2$
Block3	24 × 8,4C	24 × 8,4C	48 × 16,4C	Reduction cell	Reduction cell	Normal cell
Block4	24 × 8,4C	24 × 8,4C	48 × 16,4C	Normal cell $\times N_3$	Normal cell $\times N_3$	Normal cell $\times N_3$
Block5	12 × 4,8C	24 × 8,8C	48 × 16,8C	Reduction cell	Normal cell	Normal cell
Block6	12 × 4,8C	24 × 8,8C	48 × 16,8C	Normal cell $\times N_4$	Normal cell $\times N_4$	Normal cell $\times N_4$
Classifier	256	(1 + 2) × 256	(1 + 4) × 256	Classification cell	Classification cell	Classification cell

4.3.3 Search Space and Search Strategy

4.3.3.1 Search Space

The search space we employ is identical to that of the DARTS framework (Liu et al. 2019c). It consists of eight distinct candidate operations that the architecture search algorithm can choose from for each edge within a cell. These operations include

- skip-connect
- max-pool-3 × 3
- avg-pool-3 × 3
- sep-conv-3 × 3
- sep-conv-5 × 5
- dil-conv-3 × 3
- dil-conv-5 × 5
- zero.

In this chapter, we intentionally refrain from introducing other manually designed operations, such as the part-aware module often found in methods like Auto-ReID. Our reasoning is that a part-aware module is fundamentally a combination of these smaller, fundamental units. If the neural architecture search method is robust enough, it should be capable of discovering a similar, if not superior, functional module through the optimization process. Therefore, we prioritize trusting the search algorithm's ability to combine basic operations into effective, complex modules rather than pre-engineering such structures manually.

Algorithm 4.1 The Auto-ReID+ Algorithm

Require: The architecture parameter α and the network parameter w; the training set $\mathcal{D}_T$; the stable epochs S

Ensure: The optimal architecture $\mathcal{N}^*$

Split $\mathcal{D}_T$ into the search training set $\mathcal{D}_{train}$ and the search validation set $\mathcal{D}_{val}$
$\mathcal{N}' \leftarrow \emptyset$, $s \leftarrow 0$
while not converged **do**
  Get a batch of data from $\mathcal{D}_{train}$
  Update w via the loss in Eq. (4.10)
  Get a batch of data from $\mathcal{D}_{val}$
  Update α via the loss in Eq. (4.10)
  if the obtained architecture $\mathcal{N}(\alpha)$ is the same as $\mathcal{N}'$ **then**
    $s \leftarrow s + 1$
    if $s = S$ **then**
      $\mathcal{N}^* \leftarrow \mathcal{N}(\alpha)$
      return $\mathcal{N}^*$
    end if
  else
    $\mathcal{N}' \leftarrow \mathcal{N}(\alpha)$
    $s \leftarrow 0$
  end if
end while

4.3.3.2 Search Strategy

A common challenge in applying the standard DARTS method for neural architecture search is its tendency to produce suboptimal network architectures regarding "skip-connect" operations. Specifically, the search process can result in cells with either a complete absence of "skip-connects" or an excessive number of them. When a network architecture lacks "skip-connects", it is primarily composed of sequential convolutional layers, which can be difficult to train to convergence, often leading to poor performance. Conversely, an overabundance of "skip-connects" can severely limit the network's expressive power, also resulting in unsatisfactory performance.

DARTS+ (Liang et al. 2019b) analyzed these phenomena and proposed two early stopping criteria to mitigate them:

Criterion 1 The search procedure terminates when a single cell contains two or more "skip-connect" operations.

Criterion 2 The search procedure halts when the ranking of the architectural parameters α for the learnable operations stabilizes for a predefined number of epochs (e.g., 10 epochs).

The authors of DARTS+ experimentally demonstrated that both criteria can lead to the discovery of more effective network architectures, with Criterion 2 consistently outperforming Criterion 1. We also adopted the latter early stopping criterion within our search process. The introduction of this mechanism provides a twofold benefit: it not only enhances the efficacy of the discovered network architecture but also significantly improves the overall search efficiency. The complete procedure for our Auto-ReID+ algorithm is detailed in Algorithm 4.1. As demonstrated by our subsequent experimental results, the implementation of this early stopping mechanism successfully guides the search towards superior network architectures.

4.4 Experiments

In this section, we present the empirical evaluation of our proposed approach. The experimental setup is introduced first, covering the datasets used in Sect. 4.4.1 and implementation specifics in Sect. 4.4.2. Next, we conduct an ablation study to analyze different components of our Auto-ReID algorithm in Sect. 4.4.3. This is followed by a comparative analysis of our discovered CNN architecture against other state-of-the-art methods in Sect. 4.4.4. Finally, we provide a qualitative analysis of our results in Sect. 4.4.5.

4.4.1 Datasets and Metrics

To validate our model's performance, we performed experiments on three widely-used datasets: Market-1501, CUHK03, and MSMT17. The key characteristics of each dataset are detailed below.

Market-1501 (Zheng et al. 2015) is a large-scale ReID dataset containing 19,372 gallery images, 3,368 query images, and 12,396 training images captured from six cameras. It comprises 751 training identities and 750 test identities with no overlap. On average, each training identity has 17.2 images.

CUHK03 (Li et al. 2014) includes 1,467 unique identities and 28,192 bounding boxes. The dataset is divided into a training set with 26,264 images from 1,367 identities and a test set with 1,928 images from 100 identities. We adopt the updated data split protocol proposed by Zhong et al. (2017). CUHK03 offers two versions: one with manual labels and another with automatically detected bounding boxes. Our experiments focus on the more realistic and challenging detected version.

MSMT17 (Wei et al. 2018) is currently the largest available person ReID dataset. It features 126,441 images of 4,101 identities captured by 15 cameras. The training set consists of 32,621 bounding boxes from 1,041 identities, while the test set has 93,820 bounding boxes from 3,060 identities. The test set is further split into 11,659 query images and 82,161 gallery images. This dataset is particularly challenging due to its complex scenes and varied backgrounds, including both indoor and outdoor environments.

Evaluation Metrics. To assess the performance of our Auto-ReID model and facilitate a fair comparison with existing Re-ID approaches, we utilize two widely accepted evaluation metrics. Following the standard protocols established in literature (Zheng et al. 2015; Wei et al. 2018), we report the Cumulative Matching Characteristics (CMC) at ranks 1, 5, and 10, as well as the mean Average Precision (mAP) across all three benchmark datasets.

4.4.2 Implementation Details

4.4.2.1 Searching Configuration

During the architecture search phase, we randomly partitioned the official training set. A 50% subset of images was designated as the search training set, $\mathcal{D}_{train}$, with the remaining images forming the search validation set, $\mathcal{D}_{val}$. The foundational macro structure for the complete network, as introduced in this chapter, is detailed in Table 4.2. The network is composed of blocks, and we use the notation N_1, N_2, N_3, and N_4 to represent the number of cells within each respective block. The configuration is compactly represented by the vector $l = [N_1, N_2, N_3, N_4]$. The channel count for the initial convolutional layer is denoted as C. All specific configurations are summarized in Table 4.2. Our default search settings utilize $C = 32$ and $l = [2, 1, 1, 1]$ to identify an appropriate CNN architecture.

The DARTS search procedure is known to be computationally intensive, requiring significant GPU memory. For the person re-identification model search in this chapter, a larger batch size is advantageous for the batch hard triplet loss to effectively form more meaningful triplets. Consequently, during the search stage, the input image resolution was set to 256 × 128, differing from the 384 × 128 resolution used for the final model. This adjustment allowed us to employ a batch size of 16 × 4, where $P = 16$ and $K = 4$. The total number of epochs was set to 200, with a stabilization period of 10 epochs. We utilized momentum SGD to optimize the network weights, w, with an initial learning rate of 0.05. This learning rate was reduced to 0.0005 following a cosine annealing schedule. The momentum for SGD was 0.9. For the architecture parameters, α, the Adam optimizer was employed with an initial learning rate of 0.0005. A weight decay of 0.0005 was applied to both the SGD and Adam optimizers. The margin m for the triplet loss was set to 0.3, and the weighting factor λ between the softmax and triplet losses was set to 1. Without early stopping, the search process required approximately 2.5 days on four NVIDIA 1080Ti GPUs. However, by incorporating early stopping, the search time was reduced to about 1 day.

4.4.2.2 Training Configuration

For the training phase, we configured the network with an input image size of 384 × 128, a base channel count of $C = 64$, and the architecture vector $l = [2, 1, 1, 1]$. In line with other state-of-the-art methods, we incorporated random horizontal flipping and random erasing as data augmentation techniques. The loss function weighting parameter, λ, was set to 1. Our class-balanced data sampler was designed to construct each batch by first randomly selecting 16 distinct identities, followed by randomly sampling 4 images for each of these identities.

When training the model from scratch, the network was optimized using SGD for 350 epochs. The learning rate began at 0.065 and was reduced by a factor of 0.1 at epochs 150, 225, and 300. For fine-tuning, the network was trained for 250 epochs using the AMSGrad optimizer with an initial learning rate of 0.0015. The learning rate was decayed using a cosine annealing schedule without restarts. During the initial 10 epochs, the ImageNet pre-trained base network was frozen, and only the randomly initialized classifier was enabled for training.

4.4.3 Ablation Study

To analyze the impact of each component within our Auto-ReID+ framework, we conducted a series of extensive ablation studies on the Market-1501 dataset. The results of these experiments are presented in Table 4.3. We first compared the following configurations, all trained without ImageNet pre-training:

- When training standard ResNet-18, ResNet-34, and ResNet-50 models directly, we observed that increasing the model size and parameter count led to corresponding improvements in both mAP and rank-1 accuracy.
- We then extended the ResNet models to the multi-branch structure described in this chapter. This approach yielded a remarkable mAP of 85.0% and a rank-1 accuracy

Table 4.3 An analysis of the impact of each component within our proposed method. All CNN models were trained using the same strategy and without ImageNet pre-training. We conducted four searches for each algorithm and presented the average results. For efficiency, the search phase utilized $C = 32$ and $[N_1, N_2, N_3, N_4] = [2, 1, 1, 1]$. The final training of the discovered architecture employed $C = 64$ and $[N_1, N_2, N_3, N_4] = [2, 1, 1, 1]$. †: results as reported in Quan et al. (2019). "SC" denotes the skip-connect operation. "ES" indicates the early stopping mechanism

Architecture	mAP	R@1	R@5	R@10	Parameters (M)	FLOPs (G)	Num. SC	Search time (GPU days)
ResNet-18†	66.0	85.2	94.6	96.5	11.7	1.8	–	–
ResNet-34†	68.0	86.7	94.8	96.6	21.8	3.9	–	–
ResNet-50†	68.5	87.2	95.5	97.1	25.6	4.1	–	–
ResNet-18-Multi	81.9	93.5	97.5	98.4	36.7	12.7	–	–
ResNet-34-Multi	83.5	93.4	97.4	98.2	66.9	24.5	–	–
ResNet-50-Multi	85.0	94.3	98.0	98.4	80.6	27.3	–	–
Auto-ReID+(w/o ES, 1)	81.1	92.8	97.0	98.0	5.4	1.0	5	2.5 × 4
Auto-ReID+(w/o ES, 2)	79.0	91.1	96.0	97.3	5.5	1.0	1	2.5 × 4
Auto-ReID+(w/o ES, 3)	80.9	92.4	96.9	98.1	5.3	0.9	6	2.5 × 4
Auto-ReID+(w/o ES, 4)	82.0	92.5	96.7	97.9	5.7	1.1	3	2.5 × 4
Auto-ReID+(1)	82.7	93.6	97.5	98.3	5.8	1.1	3	0.3 × 4
Auto-ReID+(2)	85.4	94.5	97.7	98.7	5.7	1.1	4	1.2 × 4
Auto-ReID+(3)	84.5	94.1	97.8	98.5	5.5	1.0	3	1.0 × 4
Auto-ReID+(4)	83.6	93.8	97.8	98.4	5.7	1.1	4	0.9 × 4

of 94.3%, which are performance metrics on par with state-of-the-art methods. However, a significant drawback is that the multi-branch model's parameter count is roughly three times that of the single-branch version, and its FLOPs are also substantially higher. This makes such models challenging to deploy on resource-constrained mobile devices.

- We performed four independent architecture searches without implementing an early stopping mechanism, allowing each search to run for the maximum number of epochs. The resulting models maintained a low number of parameters and FLOPs, but their mAP and rank-1 scores were not as high as those of the ResNet-18-Multi model.
- Finally, we introduced an early stopping mechanism to the search process. This resulted in an architecture where the number of "skip-connect" operations in both the normal and reduction cells was consistently stabilized at 3 and 4, respectively. The best model discovered with this approach achieved a mAP of 85.4% and a rank-1 accuracy of 94.5%. This level of accuracy is exceptionally high for a model trained without any ImageNet pre-training.

4.4.3.1 The Effect of Different Configuration

We investigated the impact of different values for C and l during both the search and training phases. Due to GPU memory limitations, our search configurations for C were set to 16 and 32, while l was configured as [2, 1, 1, 1] and [3, 3, 5, 2]. For the training configurations, we used C values of 32 and 64, and the same l configurations as in the search phase. The two chosen l configurations were inspired by the ResNet architecture, which also features two distinct block count configurations, specifically [2, 2, 2, 2] and [3, 4, 6, 3].

As demonstrated in Table 4.4, an increase in the search configuration's C value led to improvements in both mAP and rank-1 accuracy. Similarly, during training, a larger C value also resulted in higher mAP and rank-1 scores. This suggests that the largest feasible C should be selected for the search phase, and a larger C can also be beneficial during training. However, it is crucial that the C values in both configurations are not excessively different, to ensure that the searched architecture's performance accurately predicts the performance of the final trained network.

In most cases, a training configuration with an l of more layers performed better than one with fewer layers. However, as the overall network structure became more refined, an l with fewer layers eventually achieved superior mAP and rank-1 scores. Considering the trade-offs between search efficiency and the size of the final network parameters, we selected an l configuration of [2, 1, 1, 1] for both the search and training phases. Ultimately, our chosen search configuration was $C = 32$ with $l =$ [2, 1, 1, 1], and our training configuration was $C = 64$ with $l =$ [2, 1, 1, 1].

4.4.3.2 The Effect of Early Stopping

We also compared the search results with and without an early stopping mechanism, while holding all other hyperparameters constant. As shown in Table 4.4, the models obtained without early stopping generally yielded inferior mAP and rank-1 scores compared to those found with it.

As previously mentioned, an excessively long search time often leads to an abundance of "skip-connect" operations in the resulting architecture, which can diminish the network's representational capacity. Conversely, we also found that if the training duration is too long, the model may end up with almost no "skip-connect" operations at all. The residual structure in deep networks, which is analogous to the "skip-connect" operation, is crucial for network convergence. Therefore, a scarcity of these connections can impede convergence and lead to unsatisfactory results.

A comparison with Table 4.4 reveals that when early stopping is applied, the number of "skip-connect" operations in the final network becomes more stable and reasonable, typically settling at either 3 or 4. Without early stopping, however, the

Table 4.4 An exploration of the impact of various configurations on both the search and training phases. In all experiments, we maintain consistent hyperparameters, with the exception of the channel size (C) and number of layers (l). None of these experiments utilize models pre-trained on ImageNet

Configurations			Rank-1	mAP	Params (M)	FLOPs (G)
Search	Train					
	C	l				
$C = 16$ $l = [2, 1, 1, 1]$	32	[2, 1, 1, 1]	90.9	79.6	2.7	0.4
		[3, 3, 5, 2]	91.3	80.3	3.7	0.7
	64	[2, 1, 1, 1]	91.6	81.2	5.6	1.0
		[3, 3, 5, 2]	91.4	80.8	7.8	1.7
$C = 16$ $l = [3, 3, 5, 2]$	32	[2, 1, 1, 1]	90.2	78.4	2.8	0.5
		[3, 3, 5, 2]	91.3	80.8	3.8	0.9
	64	[2, 1, 1, 1]	90.5	79.4	5.6	1.0
		[3, 3, 5, 2]	91.5	81.2	7.8	1.9
$C = 32$ $l = [2, 1, 1, 1]$	32	[2, 1, 1, 1]	91.9	80.2	2.9	0.4
		[3, 3, 5, 2]	92.2	80.0	3.6	0.6
	64	[2, 1, 1, 1]	94.5	85.4	5.7	1.1
		[3, 3, 5, 2]	92.9	81.6	8.0	2.0
$C = 32$ $l = [3, 3, 5, 2]$	32	[2, 1, 1, 1]	92.1	80.3	3.1	0.4
		[3, 3, 5, 2]	92.9	82.8	3.8	0.8
	64	[2, 1, 1, 1]	93.4	82.4	5.8	1.1
		[3, 3, 5, 2]	93.1	82.6	8.1	1.8

number of "skip-connect" operations fluctuates significantly, ranging from a minimum of 1 to a maximum of 6.

Furthermore, consider the time efficiency. The search process without early stopping continues until the maximum epoch is reached, costing a total of 2.5 GPU-days (since four 1080Ti GPUs were used in parallel). With early stopping, the fastest search concluded in less than 30 epochs, and most searches finished between 70 and 80 epochs. This shows that the early stopping mechanism not only improves the accuracy of the resulting network architecture but also nearly doubles the search efficiency.

4.4.4 Comparison with State-of-the-Art Methods

For our final model, we utilized the architecture discovered in the second experiment detailed in Table 4.3. Despite this, the architectures found in the other three experiments from Table 4.3 still showed significant advantages without ImageNet pre-training, which further validates the effectiveness of our algorithm's search capabilities, as demonstrated in Table 4.5. From these results, we derive the following key observations.

- When trained from scratch, our model, Auto-ReID+, achieves the best overall performance across all datasets. For example, on the Market-1501 dataset, our model's mAP surpasses OSNet, Auto-ReID, HAN, and BraidNet by 4.4%, 10.8%, 9.7%, and 15.9%, respectively. Although our model incorporates a multi-branch structure, its parameter count is approximately half of the model found by Auto-ReID, while still yielding a notable improvement in both mAP and rank-1 accuracy. Compared to OSNet, our model has double the parameters, but the substantial improvement in mAP justifies this increase. Training models from scratch for person re-identification tasks often leads to overfitting, which typically results in poor performance without the benefits of ImageNet pre-training. Our model, however, can achieve superior results with a small parameter footprint, even outperforming most models that rely on ImageNet pre-training.
- With ImageNet pre-training, our model surpasses most recently published state-of-the-art methods by a clear margin on every dataset. It is important to highlight that our model has only 5.7 million parameters, whereas most competing models use a ResNet50 backbone, which alone contains over 23.5 million parameters (excluding any additional custom modules). This demonstrates that our model is more compact and yet performs better. A comparison with MGN, a model with a similar structural concept, is particularly revealing: MGN has 70.4 million parameters, more than 12 times the number in our model. Specifically on the CUHK03 dataset, our model outperforms MGN by 8.2% and 11.3% in terms of mAP and rank-1 accuracy, respectively. This further confirms that our model is both more compact and higher-performing. Furthermore, since Auto-ReID+ is

Table 4.5 Results of SOTA methods on ReID datasets. -: not reported

Method	Venue	Backbone	Params (M)	Market1501		CUHK03		MSMT17	
				R1	mAP	R1	mAP	R1	mAP
Trained from scratch									
MobileNetV2 (Sandler et al. 2018)	CVPR'18	MobileNetV2	2.2	87.0	69.5	46.5	46.0	50.9	27.0
BraidNet (Wang et al. 2018c)	CVPR'18	BraidNet	–	83.7	69.5	–	–	–	–
HAN (Li et al. 2018)	CVPR'18	Inception	4.5	91.2	75.7	41.7	38.6	–	–
Auto-ReID (Quan et al. 2019)	ICCV'19	Auto	13.1	90.7	74.6	–	–	–	–
OSNet (Zhou et al. 2019)	ICCV'19	OSNet	2.2	93.6	81.0	57.1	54.2	71.0	43.3
Auto-ReID+(ours)	This work	Auto	5.7	94.5	85.4	69.3	67.0	77.2	53.0
Pre-trained on ImageNet									
SVDNet (Sun et al. 2017)	ICCV'17	ResNet	>23.5	82.3	62.1	41.5	37.3	–	–
PDC (Su et al. 2017)	ICCV'17	Inceptioin	>6.8	84.1	63.4	–	–	58.0	29.7
HAP2S (Yu et al. 2018)	ECCV'18	ResNet	>23.5	84.6	69.4	–	–	–	–
DPFL (Chen et al. 2017)	ICCVW'17	Inception	>6.8	88.6	72.6	40.7	37.0	–	–
DaRe (Wang et al. 2018b)	CVPR'18	DenseNet	>23.5	89.0	76.0	63.3	59.0	–	–
PNGAN (Qian et al. 2018)	ECCV'18	ResNet	>23.5	89.4	72.6	–	–	–	–
GLAD (Wei et al. 2017)	ACM MM'17	Inception	>6.8	89.9	73.9	–	–	61.4	34.0
KPM (Shen et al. 2018)	CVPR'18	ResNet	>23.5	90.1	75.3	–	–	–	–
MLFN (Chang et al. 2018)	CVPR'18	ResNeXt	32.5	90.0	74.3	52.8	47.8	–	–
FDGAN (Ge et al. 2018)	NeurIPS'18	ResNet	>23.5	90.5	77.7	–	–	–	–
DuATM (Si et al. 2018)	CVPR'18	DenseNet	>7.0	91.4	76.6	–	–	–	–
Bilinear (Suh et al. 2018)	ECCV'18	Inception	>6.8	91.7	79.6	–	–	–	–
G2G (Shen et al. 2018a)	CVPR'18	ResNet	>23.5	92.7	82.5	–	–	–	–
DeepCRF (Chen et al. 2018)	CVPR'18	ResNet	26.1	93.5	81.6	–	–	–	–
PCB (Sun et al. 2018)	ECCV'18	ResNet	27.2	93.8	81.6	63.7	57.5	68.2	40.4
SGGNN (Shen et al. 2018b)	ECCV'18	ResNet	>23.5	92.3	82.8	–	–	–	–
Mancs (Wang et al. 2018d)	ECCV'18	ResNet	>25.1	93.1	82.3	65.5	60.5	–	–
AANet (Tay et al. 2019)	CVPR'19	ResNet	>23.5	93.9	83.4	–	–	–	–
CAMA (Yang et al. 2019)	CVPR'19	ResNet	>23.5	94.7	84.5	66.6	64.2	–	–
IANet (Hou et al. 2019)	CVPR'19	ResNet	>23.5	94.4	83.1	–	–	75.5	46.8
DGNet (Zheng et al. 2019)	CVPR'19	ResNet	>23.5	94.8	86.0	65.6	61.1	77.2	52.3
MGN (Wang et al. 2018a)	ACM MM'18	ResNet	70.4	95.7	86.9	66.8	66.0	–	–
Auto-ReID (Quan et al. 2019)	ICCV'19	Auto	13.1	94.5	85.1	73.3	69.3	78.2	52.5
BagOfTricks (Luo et al. 2019)	CVPR'19	ResNet	25.6	94.5	85.9	–	–	–	–
OSNet (Zhou et al. 2019)	ICCV'19	OSNet	2.2	94.8	86.7	72.3	67.8	79.1	55.1
MHN (Chen et al. 2019b)	CVPR'19	ResNet	29.8	95.1	85.0	71.1	65.4	–	–
Auto-ReID+(ours)	This work	Auto	5.7	95.8	88.2	78.1	74.2	80.8	59.5

orthogonal to certain methods like the image generation-based DGNet, combining them could potentially lead to even greater performance improvements in person re-identification.

4.4.5 *Visualization*

4.4.5.1 The Searched Models

To provide a clearer understanding of the architectures we discovered, we have visualized them in Fig. 4.4. These architectures correspond to the four experiments detailed in Table 4.3. We present both the normal and reduction cells for each. As the figures show, the automatically searched network structures are more complex than those of manually designed networks, such as ResNet. An interesting observation is that most dilated convolutions are located exclusively within the reduction cells. This is because the primary function of the reduction cell is to down-sample the spatial dimensions of the feature map, and using dilated convolutions enhances the receptive field size, thereby boosting the network's feature representation capability. Similarly, all pooling operations are confined to the reduction cell, aligning with the design principles we adopted.

4.4.5.2 Visualization of Retrieval Results

Retrieving the correct person in the presence of distracting backgrounds can be very challenging. Figure 4.5 illustrates the retrieval results of our model on the Market-1501 dataset under such conditions. The figure shows that our model consistently retrieves the correct individuals, with most of the correct images ranked highly. Specifically, even when other people or objects are present as background interference, our model still produces accurate retrieval results. It can also successfully match a walking person to an image of them sitting on a bicycle, and a front-view image to a side-view image, further demonstrating the robustness of the features learned by our model.

However, there are still some retrieval errors. For instance, in the first row, a person wearing light blue shorts is incorrectly matched with a person carrying a light blue cloth bag at rank-4. This type of error suggests a lack of human-centric prior knowledge in the model. To address this issue, we could potentially incorporate additional data annotations.

We also present some retrieval failures in Fig. 4.6. Despite the incorrect matches, the visual similarity between the retrieved and query images is remarkably high. With the exception of some obvious mismatches, it is often difficult for a human to determine if the retrieval is correct based on appearance alone. For instance, in the first row of Fig. 4.6, the query image and the rank-1 result are so visually alike that they are nearly indistinguishable to the naked eye.

4.4.5.3 The Visualization of Feature Response Maps

We also visualized the feature response maps for each branch by summing and normalizing the feature maps across all channels. As depicted in Fig. 4.7, the feature

Fig. 4.4 Visual representation of the discovered architectures. These models correspond to the four experiments detailed in Table 4.3. Solid lines indicate one of the eight candidate operations, while dashed lines signify a predetermined operation, specifically the skip connection

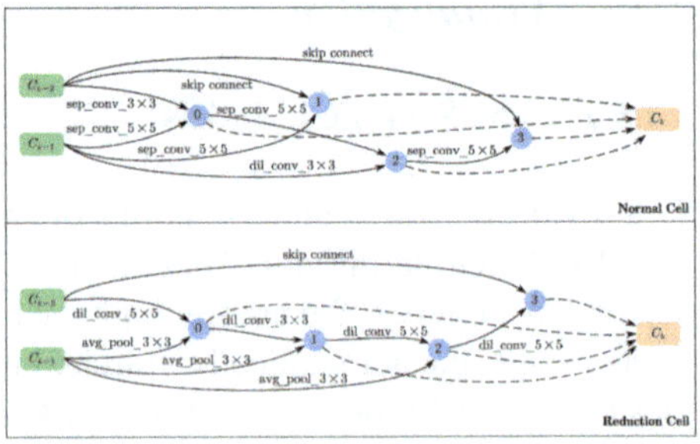

(a) run1

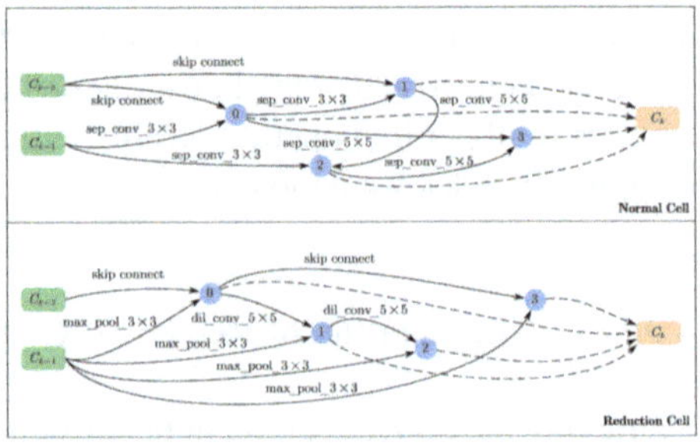

(b) run2

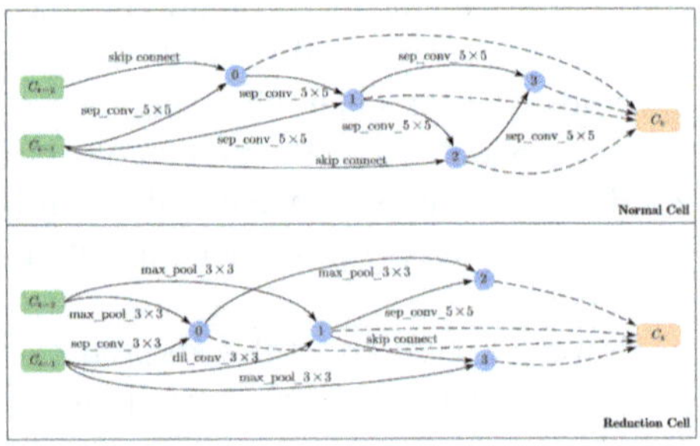

(c) run3

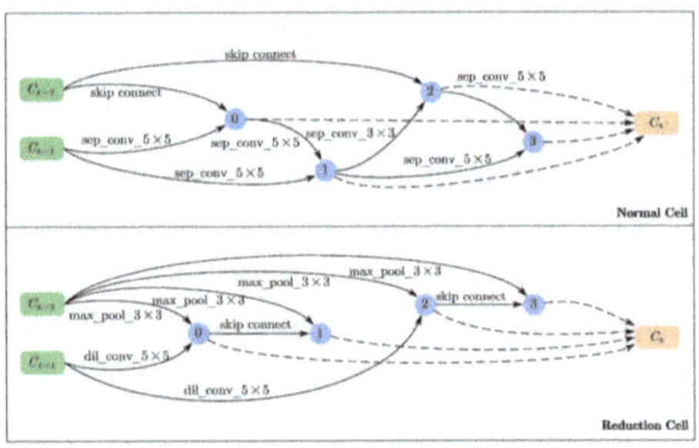

(d) run4

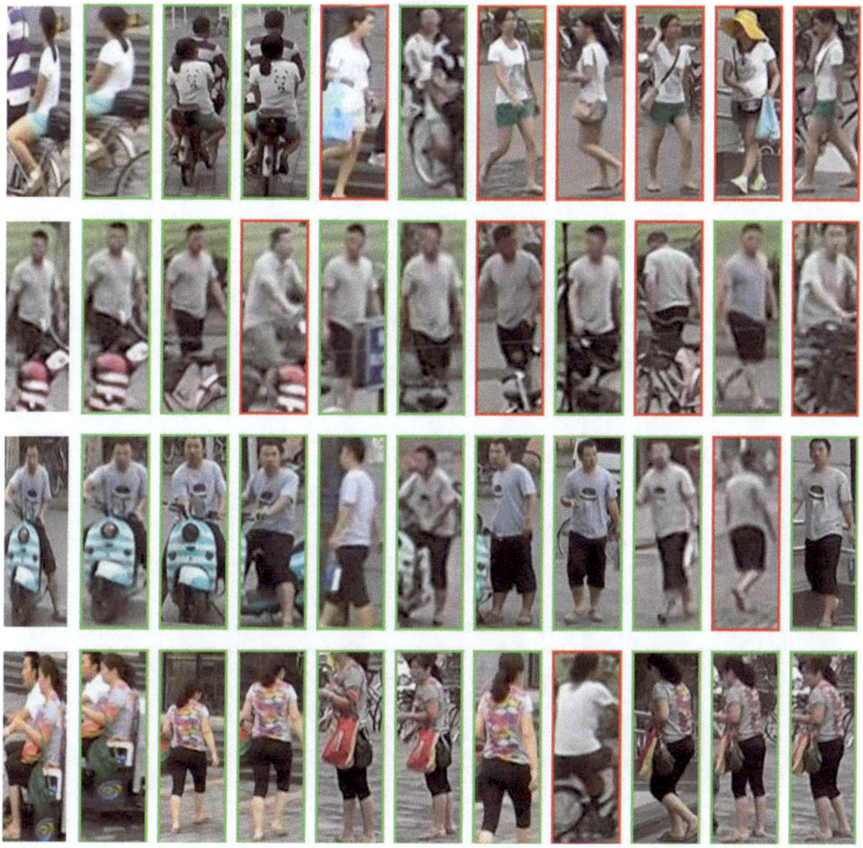

Fig. 4.5 Successful retrieval outcomes in complex scenarios on the Market-1501 dataset. The first column displays the query images, while the subsequent columns are arranged in decreasing order of similarity. A red border signifies an incorrect match, and a green border denotes a correct match

map from Branch 1 has the smallest size at 12×4. According to the theory of the effective receptive field, smaller feature maps are better suited for extracting global information. This is evident in the corresponding feature response map for Branch 1, which focuses on the entire person.

In contrast, the feature response map from Branch 2 pays more attention to specific details, primarily focusing on the head and waist regions. Branch 3, with the largest feature map, is best at capturing local information. A comparison reveals that the feature response map for Branch 3 more closely aligns with the fine-grained contours of a person, and the extracted information is highly specific. The use of a three-branch structure with varying scales facilitates a complementary fusion of global and local information, which collectively enhances the overall model's feature representation capabilities.

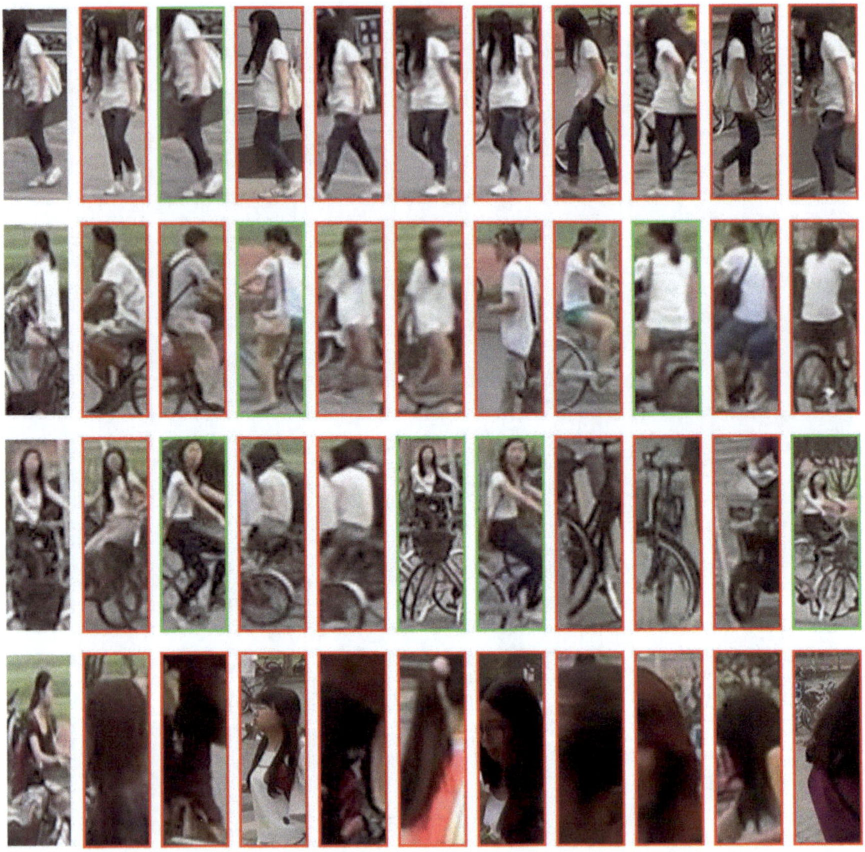

Fig. 4.6 Examples of failed retrieval outcomes on the Market-1501 dataset. The first column shows the query images, and subsequent columns are sorted in descending order of similarity. Mismatched retrievals are marked with a red border, while correct matches are indicated by a green border

4.5 Conclusion

In this chapter, we introduced Auto-ReID+, a novel approach for automatically searching for a multi-branch deep convolutional neural network specifically for person re-identification tasks. Our method successfully expands the application of Neural Architecture Search (NAS) to the person re-identification domain. By incorporating an early stopping mechanism, we not only enhanced the efficiency of the search process but also improved the performance of the resulting model. Extensive experiments have demonstrated that our model maintains a compact parameter count while achieving state-of-the-art results on several major person re-identification datasets.

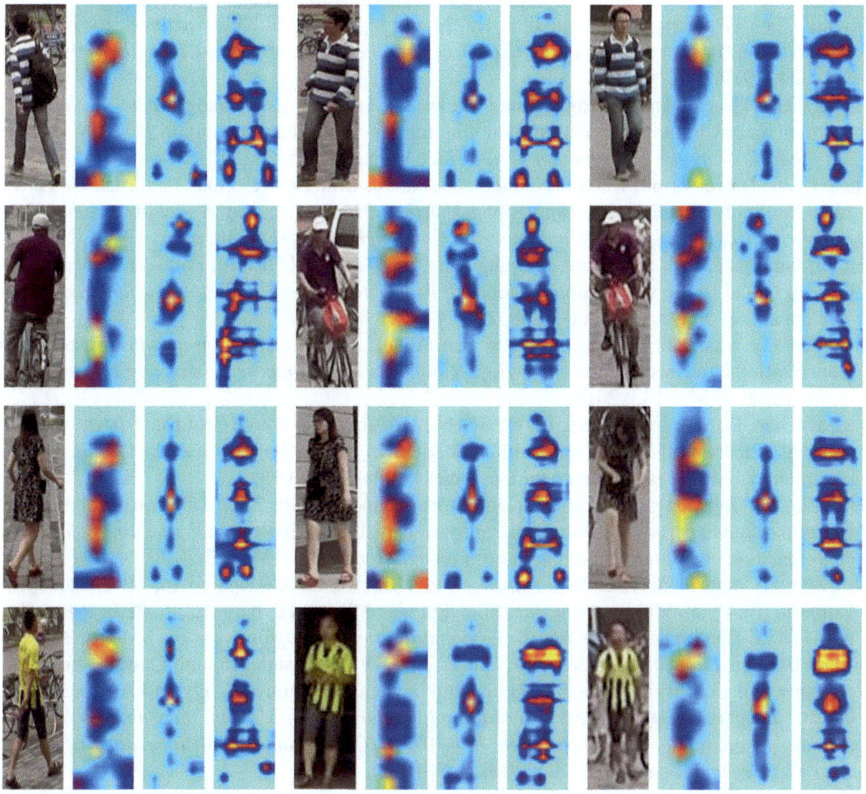

Fig. 4.7 Feature response maps derived from the output layers of each branch. Each row represents a single person, with four images grouped together: the original image, followed by the feature response maps from Branch 1, Branch 2, and Branch 3. Areas with higher intensity (redder) indicate more concentrated feature activation. For optimal viewing, please refer to the color version

References

Andrew Brock, Theodore Lim, James M. Ritchie, and Nick Weston. SMASH: one-shot model architecture search through hypernetworks. In *6th International Conference on Learning Representations, ICLR 2018, Vancouver, BC, Canada, April 30-May 3, 2018, Conference Track Proceedings*. OpenReview.net, 2018.

Antoine Yang, Pedro M. Esperança, and Fabio Maria Carlucci. NAS evaluation is frustratingly hard. In *8th International Conference on Learning Representations, ICLR 2020, Addis Ababa, Ethiopia, April 26–30, 2020*. OpenReview.net, 2020.

Barret Zoph and Quoc V Le. Neural architecture search with reinforcement learning. In *Proceedings of the International Conference on Learning Representations*, 2017.

Barret Zoph, Vijay Vasudevan, Jonathon Shlens, and Quoc V Le. Learning transferable architectures for scalable image recognition. In *Proceedings of the IEEE conference on computer vision and pattern recognition*, pages 8697–8710, 2018.

Binghui Chen, Weihong Deng, and Jiani Hu. Mixed high-order attention network for person re-identification. In *Proceedings of the IEEE International Conference on Computer Vision*, pages 371–381, 2019.

Cheng Wang, Qian Zhang, Chang Huang, Wenyu Liu, and Xinggang Wang. Mancs: A multi-task attentional network with curriculum sampling for person re-identification. In *Proceedings of the European Conference on Computer Vision (ECCV)*, pages 365–381, 2018.

Chenxi Liu, Barret Zoph, Maxim Neumann, Jonathon Shlens, Wei Hua, Li-Jia Li, Li Fei-Fei, Alan Yuille, Jonathan Huang, and Kevin Murphy. Progressive neural architecture search. In *Proceedings of the European Conference on Computer Vision (ECCV)*, pages 19–34, 2018.

Chi Su, Jianing Li, Shiliang Zhang, Junliang Xing, Wen Gao, and Qi Tian. Pose-driven deep convolutional model for person re-identification. In *Proceedings of the IEEE international conference on computer vision*, pages 3960–3969, 2017.

Chiat-Pin Tay, Sharmili Roy, and Kim-Hui Yap. Aanet: Attribute attention network for person re-identifications. In *Proceedings of the IEEE Conference on Computer Vision and Pattern Recognition*, pages 7134–7143, 2019.

Dapeng Chen, Dan Xu, Hongsheng Li, Nicu Sebe, and Xiaogang Wang. Group consistent similarity learning via deep crf for person re-identification. In *Proceedings of the IEEE conference on computer vision and pattern recognition*, pages 8649–8658, 2018.

De Cheng, Yihong Gong, Sanping Zhou, Jinjun Wang, and Nanning Zheng. Person re-identification by multi-channel parts-based cnn with improved triplet loss function. In *Proceedings of the IEEE conference on computer vision and pattern recognition*, pages 1335–1344, 2016.

Feng Zheng, Cheng Deng, Xing Sun, Xinyang Jiang, Xiaowei Guo, Zongqiao Yu, Feiyue Huang, and Rongrong Ji. Pyramidal person re-identification via multi-loss dynamic training. In *Proceedings of the IEEE Conference on Computer Vision and Pattern Recognition*, pages 8514–8522, 2019.

Gabriel Bender, Pieter-Jan Kindermans, Barret Zoph, Vijay Vasudevan, and Quoc V. Le. Understanding and simplifying one-shot architecture search. In Jennifer G. Dy and Andreas Krause, editors, *Proceedings of the 35th International Conference on Machine Learning, ICML 2018, Stockholmsmässan, Stockholm, Sweden, July 10–15, 2018*, volume 80 of *Proceedings of Machine Learning Research*, pages 549–558. PMLR, 2018.

Golnaz Ghiasi, Tsung-Yi Lin, and Quoc V Le. Nas-fpn: Learning scalable feature pyramid architecture for object detection. In *Proceedings of the IEEE Conference on Computer Vision and Pattern Recognition*, pages 7036–7045, 2019.

Guanshuo Wang, Yufeng Yuan, Xiong Chen, Jiwei Li, and Xi Zhou. Learning discriminative features with multiple granularities for person re-identification. In Susanne Boll, Kyoung Mu Lee, Jiebo Luo, Wenwu Zhu, Hyeran Byun, Chang Wen Chen, Rainer Lienhart, and Tao Mei, editors, *2018 ACM Multimedia Conference on Multimedia Conference, MM 2018, Seoul, Republic of Korea, October 22–26, 2018*, pages 274–282. ACM, 2018.

Hanwen Liang, Shifeng Zhang, Jiacheng Sun, Xingqiu He, Weiran Huang, Kechen Zhuang, and Zhenguo Li. Darts+: Improved differentiable architecture search with early stopping. *arXiv preprint* arXiv:1909.06035, 2019.

Hanwen Liang, Shifeng Zhang, Jiacheng Sun, Xingqiu He, Weiran Huang, Kechen Zhuang, and Zhenguo Li. DARTS+: improved differentiable architecture search with early stopping. *CoRR*, abs/1909.06035, 2019.

Hanxiao Liu, Karen Simonyan, and Yiming Yang. DARTS: differentiable architecture search. In *7th International Conference on Learning Representations, ICLR 2019, New Orleans, LA, USA, May 6–9, 2019*, 2019.

Hao Luo, Youzhi Gu, Xingyu Liao, Shenqi Lai, and Wei Jiang. Bag of tricks and a strong baseline for deep person re-identification. In *Proceedings of the IEEE Conference on Computer Vision and Pattern Recognition Workshops*, pages 0–0, 2019.

Jianlou Si, Honggang Zhang, Chun-Guang Li, Jason Kuen, Xiangfei Kong, Alex C Kot, and Gang Wang. Dual attention matching network for context-aware feature sequence based person re-identification. In *Proceedings of the IEEE Conference on Computer Vision and Pattern Recognition*, pages 5363–5372, 2018.

Jiawei Liu, Zheng-Jun Zha, Di Chen, Richang Hong, and Meng Wang. Adaptive transfer network for cross-domain person re-identification. In *IEEE Conference on Computer Vision and Pattern Recognition, CVPR 2019, Long Beach, CA, USA, June 16–20, 2019*, pages 7202–7211. Computer Vision Foundation/IEEE, 2019.

Jiawei Liu, Zheng-Jun Zha, Xuejin Chen, Zilei Wang, and Yongdong Zhang. Dense 3d-convolutional neural network for person re-identification in videos. *ACM Trans. Multim. Comput. Commun. Appl.*, 15(1s):8:1–8:19, 2019.

Jiawei Liu, Zheng-Jun Zha, Q. I. Tian, Dong Liu, Ting Yao, Qiang Ling, and Tao Mei. Multi-scale triplet CNN for person re-identification. In Alan Hanjalic, Cees Snoek, Marcel Worring, Dick C. A. Bulterman, Benoit Huet, Aisling Kelliher, Yiannis Kompatsiaris, and Jin Li, editors, *Proceedings of the 2016 ACM Conference on Multimedia Conference, MM 2016, Amsterdam, The Netherlands, October 15–19, 2016*, pages 192–196. ACM, 2016.

Kaiyang Zhou, Yongxin Yang, Andrea Cavallaro, and Tao Xiang. Omni-scale feature learning for person re-identification. In *Proceedings of the IEEE International Conference on Computer Vision*, pages 3702–3712, 2019.

Liang Zheng, Liyue Shen, Lu Tian, Shengjin Wang, Jingdong Wang, and Qi Tian. Scalable person re-identification: A benchmark. In *Proceedings of the IEEE international conference on computer vision*, pages 1116–1124, 2015.

Longhui Wei, Shiliang Zhang, Hantao Yao, Wen Gao, and Qi Tian. Glad: Global-local-alignment descriptor for pedestrian retrieval. In *Proceedings of the 25th ACM international conference on Multimedia*, pages 420–428, 2017.

Longhui Wei, Shiliang Zhang, Wen Gao, and Qi Tian. Person transfer gan to bridge domain gap for person re-identification. In *Proceedings of the IEEE conference on computer vision and pattern recognition*, pages 79–88, 2018.

Mark Sandler, Andrew Howard, Menglong Zhu, Andrey Zhmoginov, and Liang-Chieh Chen. Mobilenetv2: Inverted residuals and linear bottlenecks. In *Proceedings of the IEEE conference on computer vision and pattern recognition*, pages 4510–4520, 2018.

Rui Yu, Zhiyong Dou, Song Bai, Zhaoxiang Zhang, Yongchao Xu, and Xiang Bai. Hard-aware point-to-set deep metric for person re-identification. In *Proceedings of the European Conference on Computer Vision (ECCV)*, pages 188–204, 2018.

Ruibing Hou, Bingpeng Ma, Hong Chang, Xinqian Gu, Shiguang Shan, and Xilin Chen. Interaction-and-aggregation network for person re-identification. In *Proceedings of the IEEE Conference on Computer Vision and Pattern Recognition*, pages 9317–9326, 2019.

Ruijie Quan, Xuanyi Dong, Yu Wu, Linchao Zhu, and Yi Yang. Auto-reid: Searching for a part-aware convnet for person re-identification. In *2019 IEEE/CVF International Conference on Computer Vision, ICCV 2019, Seoul, Korea (South), October 27- November 2, 2019*, pages 3749–3758. IEEE, 2019.

Wei Li, Rui Zhao, Tong Xiao, and Xiaogang Wang. Deepreid: Deep filter pairing neural network for person re-identification. In *Proceedings of the IEEE conference on computer vision and pattern recognition*, pages 152–159, 2014.

Wei Li, Xiatian Zhu, and Shaogang Gong. Harmonious attention network for person re-identification. In *Proceedings of the IEEE conference on computer vision and pattern recognition*, pages 2285–2294, 2018.

Wenjie Luo, Yujia Li, Raquel Urtasun, and Richard Zemel. Understanding the effective receptive field in deep convolutional neural networks. In *Advances in neural information processing systems*, pages 4898–4906, 2016.

Wenjie Yang, Houjing Huang, Zhang Zhang, Xiaotang Chen, Kaiqi Huang, and Shu Zhang. Towards rich feature discovery with class activation maps augmentation for person re-identification. In *Proceedings of the IEEE Conference on Computer Vision and Pattern Recognition*, pages 1389–1398, 2019.

Xiangxiang Chu, Tianbao Zhou, Bo Zhang, and Jixiang Li. Fair DARTS: eliminating unfair advantages in differentiable architecture search. *CoRR*, abs/1911.12126, 2019.

Xiaobin Chang, Timothy M Hospedales, and Tao Xiang. Multi-level factorisation net for person re-identification. In *Proceedings of the IEEE Conference on Computer Vision and Pattern Recognition*, pages 2109–2118, 2018.

Xuelin Qian, Yanwei Fu, Tao Xiang, Wenxuan Wang, Jie Qiu, Yang Wu, Yu-Gang Jiang, and Xiangyang Xue. Pose-normalized image generation for person re-identification. In *Proceedings of the European conference on computer vision (ECCV)*, pages 650–667, 2018.

Yan Wang, Lequn Wang, Yurong You, Xu Zou, Vincent Chen, Serena Li, Gao Huang, Bharath Hariharan, and Kilian Q Weinberger. Resource aware person re-identification across multiple resolutions. In *Proceedings of the IEEE Conference on Computer Vision and Pattern Recognition*, pages 8042–8051, 2018.

Yanbei Chen, Xiatian Zhu, and Shaogang Gong. Person re-identification by deep learning multi-scale representations. In *Proceedings of the IEEE International Conference on Computer Vision Workshops*, pages 2590–2600, 2017.

Yantao Shen, Hongsheng Li, Shuai Yi, Dapeng Chen, and Xiaogang Wang. Person re-identification with deep similarity-guided graph neural network. In *Proceedings of the European conference on computer vision (ECCV)*, pages 486–504, 2018.

Yantao Shen, Hongsheng Li, Tong Xiao, Shuai Yi, Dapeng Chen, and Xiaogang Wang. Deep group-shuffling random walk for person re-identification. In *Proceedings of the IEEE conference on computer vision and pattern recognition*, pages 2265–2274, 2018.

Yantao Shen, Tong Xiao, Hongsheng Li, Shuai Yi, and Xiaogang Wang. End-to-end deep kronecker-product matching for person re-identification. In *Proceedings of the IEEE Conference on Computer Vision and Pattern Recognition*, pages 6886–6895, 2018.

Yicheng Wang, Zhenzhong Chen, Feng Wu, and Gang Wang. Person re-identification with cascaded pairwise convolutions. In *Proceedings of the IEEE Conference on Computer Vision and Pattern Recognition*, pages 1470–1478, 2018.

Yifan Sun, Liang Zheng, Weijian Deng, and Shengjin Wang. Svdnet for pedestrian retrieval. In *Proceedings of the IEEE International Conference on Computer Vision*, pages 3800–3808, 2017.

Yifan Sun, Liang Zheng, Yi Yang, Qi Tian, and Shengjin Wang. Beyond part models: Person retrieval with refined part pooling (and a strong convolutional baseline). In *Proceedings of the European Conference on Computer Vision (ECCV)*, pages 480–496, 2018.

Yixiao Ge, Zhuowan Li, Haiyu Zhao, Guojun Yin, Shuai Yi, Xiaogang Wang, et al. Fd-gan: Pose-guided feature distilling gan for robust person re-identification. In *Advances in neural information processing systems*, pages 1222–1233, 2018.

Yuhui Xu, Lingxi Xie, Xiaopeng Zhang, Xin Chen, Guo-Jun Qi, Qi Tian, and Hongkai Xiong. PC-DARTS: partial channel connections for memory-efficient architecture search. In *8th International Conference on Learning Representations, ICLR 2020, Addis Ababa, Ethiopia, April 26–30, 2020*. OpenReview.net, 2020.

Yukang Chen, Tong Yang, Xiangyu Zhang, Gaofeng Meng, Chunhong Pan, and Jian Sun. Detnas: Neural architecture search on object detection. *CoRR*, abs/1903.10979, 2019.

Yumin Suh, Jingdong Wang, Siyu Tang, Tao Mei, and Kyoung Mu Lee. Part-aligned bilinear representations for person re-identification. In *Proceedings of the European Conference on Computer Vision (ECCV)*, pages 402–419, 2018.

Zhedong Zheng, Xiaodong Yang, Zhiding Yu, Liang Zheng, Yi Yang, and Jan Kautz. Joint discriminative and generative learning for person re-identification. In *Proceedings of the IEEE Conference on Computer Vision and Pattern Recognition*, pages 2138–2147, 2019.

Zhun Zhong, Liang Zheng, Donglin Cao, and Shaozi Li. Re-ranking person re-identification with k-reciprocal encoding. In *Proceedings of the IEEE Conference on Computer Vision and Pattern Recognition*, pages 1318–1327, 2017.

Chapter 5
Dynamic Margin-Based Softmax Loss Function Search for Person Re-identification

5.1 Introduction

The increasing importance of public safety has positioned person re-identification (ReID) as a pivotal research area in computer vision, serving as a critical component in contemporary video surveillance systems (Di et al. 2019; Zheng et al. 2019a; Hou et al. 2019; Zhang et al. 2021a; Wanyin et al. 2021; Fan et al. 2019). The core task of ReID is to identify and retrieve a specific person of interest from a collection of images or videos, often referred to as a gallery set, captured by a network of non-overlapping cameras. Recent advances in deep learning have led to the development of numerous models based on deep convolutional neural networks (DCNNs), which have dramatically enhanced performance. A well-trained DCNN learns to produce feature representations where the feature vectors for the same individual are clustered closely together, while those for different people are pushed far apart, as determined by a given similarity metric.

A key challenge in training deep neural networks for ReID is designing an effective loss function that enhances the discriminative power of the extracted features. Existing models can be broadly categorized into three groups based on their loss function strategies. **Classification-based models** frame the ReID problem as a classification task, directly using the cross-entropy loss function to classify features learned by DCNNs. However, as noted in several studies (Fan et al. 2019; Wen et al. 2016; Liu et al. 2017; Deng et al. 2019; Luo et al. 2019; Sun et al. 2020), the features learned this way often achieve simple separability rather than true discriminative power. This is problematic for ReID, where identities in the training set are distinct from those in the testing set. **Metric learning-based models** typically employ loss functions from metric learning (Hermans et al. 2017), such as contrastive or triplet loss. While these functions are theoretically well-suited for ReID, their practical implementation can be computationally expensive. To mitigate this, a carefully designed sample mining strategy is often required (Wu et al. 2017; Roth et al. 2020). **Hybrid models** that combine both classification and metric learning losses have become a popular approach

H. Gu et al., *Automated Machine Learning for Person Re-Identification*,
Intelligent Perception and Information Processing,
https://doi.org/10.1007/978-981-95-3433-3_5

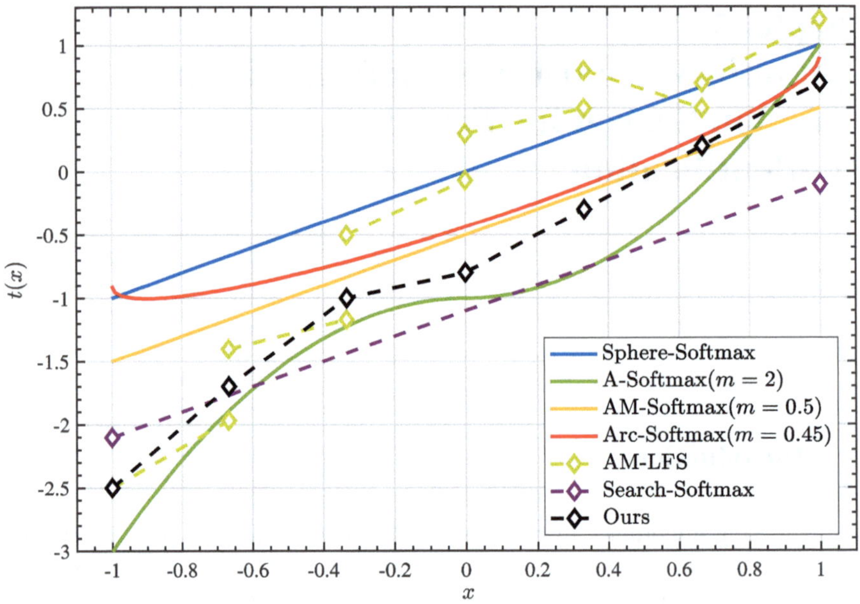

Fig. 5.1 Comparison of $t(x)$ curves across various margin-based softmax loss functions. The solid line depicts the manually crafted $t(x)$, while the dotted line illustrates the search domain defined by AutoML

for training ReID models. The most common combination involves cross-entropy and triplet loss. However, applying both losses to the same feature can lead to conflicting optimization goals, which may compromise the generalization of the final learned features (Luo et al. 2019; Zheng et al. 2019b). Methods like MGN (Wang et al. 2018a) and StrongBaseline (Luo et al. 2019) address this by carefully tailoring the network architecture to apply the two losses to different features. Furthermore, Pyramid (Zheng et al. 2019b) introduced an algorithm for dynamically adjusting the weights of these loss functions. These examples highlight that significant manual effort is often required to tune hyperparameters, design network architectures, or optimize sample mining strategies.

Furthermore, metrics like mean Average Precision (mAP) are non-differentiable and non-decomposable, as they depend on the collective statistics of a set of predictions and ground-truth labels rather than individual samples. Therefore, it is necessary to engineer a differentiable loss function that can serve as a suitable proxy for the target metric (Huang et al. 2019). However, no matter how meticulously a loss function is crafted, it is unlikely to perfectly capture the objective metric throughout the entire training process.

Recently, an increasing number of methods based on AutoML have emerged. In particular, automated approaches have surpassed manual design in areas like data augmentation (Cubuk et al. 2019; Lim et al. 2019; Zhang et al. 2020; Tian et al. 2020) and network architecture search (Zoph and Le 2017; Pham et al. 2018; Liu et al. 2019;

Zhang et al. 2021b). However, the design of loss functions largely remains a manual process.

AM-LFS (Li et al. 2019) and Search-Softmax (SS) (Wang et al. 2020) represent two efforts to apply AutoML to loss function search. These methods aim to automatically design a function, $t(x)$, guided by a target metric on a validation dataset, thereby implicitly optimizing the target metric throughout training. AM-LFS uses a piecewise linear function to approximate $t(x)$, where the slopes and intercepts are independently optimized. As noted by SS (Wang et al. 2020), this search space is complex and unstable, a finding that our own experiments also confirm. In contrast, SS uses a search space inspired by AM-Softmax (Wang et al. 2018b), which is too simplistic to capture a wide variety of loss function forms. Both search spaces fail to guarantee some fundamental properties of $t(x)$.

In this chapter, we propose LFS-ReID, an AutoML approach for loss function search specifically for person re-identification. Our method introduces a more reasonable search space. We first conduct a thorough analysis of margin-based softmax loss from three perspectives–gradient, feature, and optimization—to derive four key properties. We then carefully design a sampling distribution based on non-independent truncated Gaussian distributions. This distribution allows us to sample loss functions that inherently conform to the four derived properties. Furthermore, we adopt a reinforcement learning-based method within a bi-level optimization framework to alternately optimize the parameters of the sampling distribution and the neural network. The inner optimization loop uses stochastic gradient descent (SGD) to train the neural network with the currently sampled loss function. The outer optimization loop employs a policy gradient to optimize the parameters of the sampling distribution, aiming for a higher reward (e.g., mAP) on the validation dataset from the trained model. Finally, upon completion of the LFS-ReID training, the network parameters can be used directly for evaluation without any re-training. A key advantage of our method is that it optimizes the loss function without modifying the backbone network architecture, making it easily applicable to almost all existing ReID models. The main contributions of this work are summarized as follows:

- We present an analysis of margin-based softmax loss, concluding four key properties from the perspectives of gradient, feature, and optimization.
- We design a novel sampling distribution based on non-independent truncated Gaussian distributions, ensuring that the sampled loss functions adhere to these four properties.
- Our proposed method achieves state-of-the-art performance on three commonly used datasets.

5.2 Related Works

5.2.1 Person Re-identification

Person re-identification is a prominent research area in computer vision, with current algorithms achieving significant success using deep learning techniques (Zheng et al. 2019a; Wang et al. 2018a; Chen et al. 2019; Li et al. 2018; Sun et al. 2018; Wu et al. 2020). These models can be generally classified into two categories based on their feature extraction approach: those relying on global features and those using local features. The former directly utilizes global features from backbone networks like ResNet50 (He et al. 2016) or InceptionV3 (Szegedy et al. 2016). This approach, however, is susceptible to overfitting (Sun et al. 2017), which can limit overall performance.

Models that focus on local features can learn more detailed information, often leading to better performance. For instance, the PCB model proposed by Sun et al. (2018) uses prior knowledge of the human body structure to vertically and uniformly partition the backbone feature map, from which it extracts local features. This simple yet effective design has been widely adopted in subsequent researches (Zheng et al. 2019b; Wang et al. 2018a; Yang et al. 2019). A notable example is MGN (Wang et al. 2018a), which uses a multi-branch network architecture to integrate both global and local features.

Our method primarily focuses on optimizing the loss function without making any changes to the backbone network architecture. This design choice makes our approach easily adaptable to nearly all existing ReID models. We validate the effectiveness of our loss function on four common backbone network architectures in the subsequent experiments.

5.2.2 Loss Function

Loss functions are crucial for improving the discriminative power of learned features. Standard softmax loss tends to produce features that are merely separable rather than highly discriminative. To address this, several margin-based softmax loss functions have been proposed in recent years to enhance feature discrimination (Fan et al. 2019; Wen et al. 2016; Liu et al. 2017; Deng et al. 2019; Luo et al. 2019; Sun et al. 2020).

All the aforementioned loss functions are manually designed, a process that requires significant effort. To automate this, Li et al. (2019)vintroduced AM-LFS, an AutoML method for searching for the specific form of the function $t(x)$. However, the search space of AM-LFS has been criticized as being complex and unstable. Wang et al. (2020) further analyzed margin-based softmax loss and proposed that the key to a good loss function is reducing the softmax probability. Based on this insight, they introduced a new loss function search method called Search-Softmax (SS).

Compared to the search spaces of these two AutoML-based methods, ours offers a stable middle ground between AM-LFS and SS, maintaining a balance between stability and diversity of sampled loss functions. We demonstrate the advantages of our method both in its ability to express $t(x)$ and through subsequent experiments.

5.3 Methodology

First, we review several standard loss functions in Sect. 5.3.1. Subsequently, we deduce four crucial characteristics of margin-based softmax loss, analyzing them from the three viewpoints of gradient, feature, and optimization in Sect. 5.3.2. Follow-

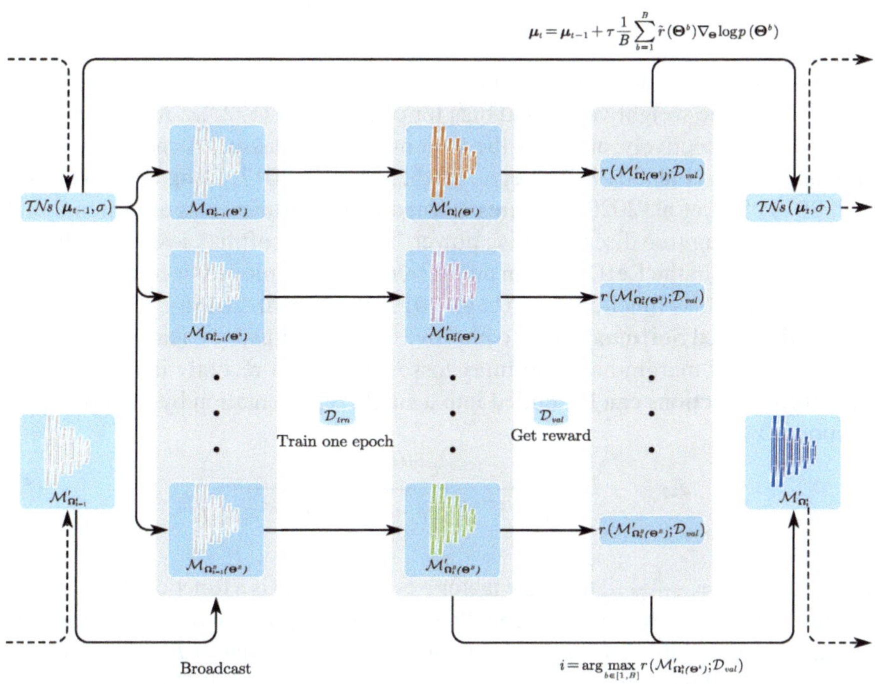

Fig. 5.2 At epoch t, the LFS-ReID optimization procedure samples B unique sets of **meta-parameters** from a truncated normal distribution, $\mathcal{TN}s(\mu_{t-1}, \sigma)$. These sets, denoted $\Theta^1, \Theta^2, ..., \Theta^B$, are subsequently allocated to B models, which are trained concurrently for a single epoch. Each model is initialized from a shared baseline, $\mathcal{M}'_{\Omega^i_{t-1}}$. Upon completion of training, the models are assessed on a validation dataset to compute their respective **scores**. Ultimately, these scores are leveraged to update the parameters of the meta-parameter sampling distribution, using the equation in Eq. (5.21)

ing this, we formulate the sampling distribution utilizing non-independent truncated Gaussian distributions in Sect. 5.3.3, guaranteeing that the resulting loss function conforms to the four properties detailed in Sect. 5.3.2. Finally, we employ a reinforcement learning framework to dynamically refine the parameters of the sampling distribution in Sect. 5.3.4. The complete optimization process for our LFS-ReID approach is illustrated in Fig. 5.2.

5.3.1 Preliminary Knowledge

Softmax Loss. The most prevalent softmax loss, commonly applied in classification tasks, is expressed as:

$$\mathcal{L}_1 = -\log\left(\frac{e^{\boldsymbol{w}_y^{\mathrm{T}}\boldsymbol{x}+b_y}}{e^{\boldsymbol{w}_y^{\mathrm{T}}\boldsymbol{x}+b_y} + \sum_{k\neq y}^{K} e^{\boldsymbol{w}_k^{\mathrm{T}}\boldsymbol{x}+b_k}}\right), \tag{5.1}$$

where the input feature vector is denoted as $\boldsymbol{x} \in \mathbb{R}^d$ and the true label is $y \in \{1, 2, ..., K\}$. The weight vector and bias for class k ($k \in \{1, 2, ..., K\}$) are $\boldsymbol{w}_k \in \mathbb{R}^d$ and $b_k \in \mathbb{R}$, respectively, and K is the total number of classes. As noted by numerous studies (Fan et al. 2019; Wen et al. 2016; Liu et al. 2017; Deng et al. 2019; Luo et al. 2019; Sun et al. 2020), features learned using softmax loss are often merely separable, lacking true discriminative power. Therefore, softmax loss by itself is not highly effective in the ReID domain and is frequently employed in conjunction with metric learning loss functions (Luo et al. 2019; Wang et al. 2018a).

Margin-based Softmax Loss. To improve the discriminative capability of learned features, several margin-based softmax losses have been recently introduced. This class of loss functions can be unified into a single representation by incorporating a function $t(x)$:

$$\mathcal{L}_2 = -\log\left(\frac{e^{s\cdot t\left(\cos\theta_{\boldsymbol{w}_y,\boldsymbol{x}}\right)}}{e^{s\cdot t\left(\cos\theta_{\boldsymbol{w}_y,\boldsymbol{x}}\right)} + \sum_{k\neq y}^{K} e^{s\cdot\cos\theta_{\boldsymbol{w}_k,\boldsymbol{x}}}}\right). \tag{5.2}$$

where $\cos\theta_{\boldsymbol{w}_k,\boldsymbol{x}} = \frac{\boldsymbol{w}_k^{\mathrm{T}}\boldsymbol{x}}{\|\boldsymbol{w}_k\|\|\boldsymbol{x}\|}$, the scale factor $s > 0$, and $t(x)$ is a function defined on the domain $[-1, 1]$. A few of these manually designed margin-based softmax loss functions are listed in Table 5.1. For example, setting $t(x) = x$ makes $\mathcal{L}_2$ degenerate into sphere-softmax loss (Fan et al. 2019). A-softmax loss (Liu et al. 2017) is motivated by $t(x) = \cos(m \cdot \arccos(x))$, where $m \geqslant 1$ is an integer. AM-softmax loss (Wang et al. 2018c, b) is obtained by setting $t(x) = x - m$, where $m \geqslant 0$. Finally, Arc-softmax loss (Deng et al. 2019) can be derived with $t(x) = \cos(\arccos(x) + m)$, where $m \geqslant 0$.

AutoML Loss. All the methods introduced above are based on a manually designed $t(x)$, which is a costly, time-consuming, and often suboptimal process. Li et al. (2019) observed that the main role of margin-based softmax loss is to use

Table 5.1 Variants of margin-based softmax loss. "P1", "P2", "P3", and "P4" refer to "Property 1", "Property 2", "Property 3", and "Property 4", respectively. "Search Space" denotes the dimensionality of the search space's parameters. "Sampling Distribution" shows the distribution from which each loss samples

Loss	$t(x)$	P1	P2	P3	P4	Search space	Sampling distribution
Sphere-Softmax (Fan et al. 2019)	x	✓	✗	✓	✓	–	–
A-Softmax (Liu et al. 2017)	$\cos(m \arccos(x))$	✓	✓	✓	✓	–	–
AM-Softmax (Wang et al. 2018c, b)	$x - m$	✓	✗	✓	✓	–	–
Arc-Softmax (Deng et al. 2019)	$\cos(\arccos(x) + m)$	✓	✓	✓	✓	–	–
AM-LFS (Li et al. 2019)	$a_i x + b_i, i = 1, 2, ..., M$	✗	✓	✗	✗	$2M$	$a_i \sim \mathcal{N}(\mu_{a_i}, \sigma^2), b_i \sim \mathcal{N}(\mu_{b_i}, \sigma^2)$
SS (Wang et al. 2020)	$x - a$	✓	✗	✓	✗	1	$a \sim \mathcal{N}(\mu_a, \sigma^2), a \geqslant 0$
Ours	$pwl(x; a_0, a_1, ..., a_M)$	✓	✓	✓	✓	$M + 1$	$[a_0, a_1, ..., a_M]^{\mathrm{T}} \sim \mathcal{TN}s(\boldsymbol{\mu}, \sigma)$

its gradient, $t'(\cos\theta_{w_y,x})$, to control the relative significance of intra-class vs. inter-class distance during training. They proposed an AutoML method for loss function search, named AM-LFS, to automatically find the specific functional form of $t(x)$.[1] Specifically, they defined $t(x)$ as a piecewise linear function over the interval $[-1, 1]$:

$$t(x) = a_i x + b_i, x \in [-1 + \frac{2i}{M}, -1 + \frac{2(i+1)}{M}]. \tag{5.3}$$

AM-LFS partitions the interval $[-1, 1]$ into M equal segments. Within each i-th sub-interval, the slope a_i and intercept b_i of the linear function are independently drawn from a Gaussian distribution. This independent sampling often results in $t(x)$ being a discontinuous, piecewise linear function, as shown in Fig. 5.1.

The search space of AM-LFS is intricate and lacks stability. Wang et al. (2020) conducted a deeper analysis of margin-based softmax loss, suggesting that the primary challenge in loss function design is minimizing the softmax probability. They introduced Search-Softmax (SS), a new loss function search approach that substantially simplifies the search space of AM-LFS by only adjusting a':

$$a' = 1 - e^{s[\cos\theta_{w_y,x} - t(\cos\theta_{w_y,x})]}, a' \leqslant 0. \tag{5.4}$$

In SS, a' is treated as an optimized hyperparameter. By letting $a = \ln(1 - a')/s$ and $x = \cos\theta_{w_y,x}$, the optimization of a' can be converted to the optimization of a in $t(x)$:

$$t(x) = x - a, x \in [-1, 1], a \geqslant 0. \tag{5.5}$$

The search space of SS is significantly simplified, requiring the optimization of only one parameter, a, while still satisfying the condition of reducing the softmax probability.

However, the functional form in the search space of SS is too simple, which limits its performance (as shown in subsequent experiments). Conversely, the search space of AM-LFS is too arbitrary, leading to instability (as also shown in subsequent experiments). Therefore, we believe that the functions in the search space must possess a certain level of diversity while also adhering to necessary properties to avoid being overly arbitrary.

---

[1] In AM-LFS, another piecewise linear function $\tau(p_y^t)$ is also optimized on p_y^t in Eq. (5.8). The original intention of $\tau(p_y^t)$ in AM-LFS is to simulate focal loss to solve the problem of class imbalance. We use a class-balance sampling method to construct the batch, so we no longer consider the optimization of $\tau(p_y^t)$.

5.3.2 Analysis of Margin-Based Softmax Loss

First, we will analyze the properties that $t(x)$ should possess from a gradient perspective. According to Eq. (5.2), the gradients of $\mathcal{L}_2$ with respect to $\cos\theta_{\boldsymbol{w}_y,\boldsymbol{x}}$ and $\cos\theta_{\boldsymbol{w}_k,\boldsymbol{x}}\,(k \neq y)$ are as follows:

$$\frac{\partial \mathcal{L}_2}{\partial \cos\theta_{\boldsymbol{w}_y,\boldsymbol{x}}} = -s \cdot t'\left(\cos\theta_{\boldsymbol{w}_y,\boldsymbol{x}}\right) \cdot \left(1 - p_y^t\right), \tag{5.6}$$

$$\frac{\partial \mathcal{L}_2}{\partial \cos\theta_{\boldsymbol{w}_k,\boldsymbol{x}}} = s \cdot p_k^t,\ k \neq y, \tag{5.7}$$

where

$$p_y^t = \frac{e^{s \cdot t\left(\cos\theta_{\boldsymbol{w}_y,\boldsymbol{x}}\right)}}{e^{s \cdot t\left(\cos\theta_{\boldsymbol{w}_y,\boldsymbol{x}}\right)} + \sum_{k\neq y}^{K} e^{s\cdot\cos\theta_{\boldsymbol{w}_k,\boldsymbol{x}}}}, \tag{5.8}$$

$$p_k^t = \frac{e^{s\cdot\cos\theta_{\boldsymbol{w}_k,\boldsymbol{x}}}}{e^{s \cdot t\left(\cos\theta_{\boldsymbol{w}_y,\boldsymbol{x}}\right)} + \sum_{j\neq y}^{K} e^{s\cdot\cos\theta_{\boldsymbol{w}_j,\boldsymbol{x}}}},\ k \neq y. \tag{5.9}$$

As training progresses, the overall trend of $\cos\theta_{\boldsymbol{w}_y,\boldsymbol{x}}$ increases, while the overall trend of $\cos\theta_{\boldsymbol{w}_k,\boldsymbol{x}}\,(k \neq y)$ decreases. Therefore, from a macro trend, the following properties hold:

$$\frac{\partial \mathcal{L}_2}{\partial \cos\theta_{\boldsymbol{w}_y,\boldsymbol{x}}} \leqslant 0,\ \frac{\partial \mathcal{L}_2}{\partial \cos\theta_{\boldsymbol{w}_k,\boldsymbol{x}}} \geqslant 0,\ k \neq y.$$

We derive the initial property by combining it with Eq. (5.6).

Property 1. $t'(x) \geqslant 0,\ x \in [-1, 1]$.
Based on Eqs. (5.6) and (5.7), we can define the relative gradientr_y^k (Li et al. 2019) as:

$$r_y^k = \frac{\left\|\frac{\partial \mathcal{L}_2}{\partial \cos\theta_{\boldsymbol{w}_y,\boldsymbol{x}}}\right\|}{\left\|\frac{\partial \mathcal{L}_2}{\partial \cos\theta_{\boldsymbol{w}_k,\boldsymbol{x}}}\right\|} = \frac{t'\left(\cos\theta_{\boldsymbol{w}_y,\boldsymbol{x}}\right)\cdot\left(1 - p_y^t\right)}{p_k^t}. \tag{5.10}$$

For a simplified case, such as binary classification, Eq. (5.10) can be further reduced to:

$$r_y^k = t'\left(\cos\theta_{\boldsymbol{w}_y,\boldsymbol{x}}\right).$$

From the formula above, we can see that $t'(x)$ controls the magnitude of the relative gradient. The relative gradient, in turn, reflects the ratio between the positive gradient, $\frac{\partial \mathcal{L}_2}{\partial \cos\theta_{\boldsymbol{w}_y,\boldsymbol{x}}}$, and the negative gradient, $\frac{\partial \mathcal{L}_2}{\partial \cos\theta_{\boldsymbol{w}_k,\boldsymbol{x}}}$. As analyzed in the paper on circle loss (Sun et al. 2020), samples of varying difficulty should have different relative gradient magnitudes. This leads us to another key property.

Property 2. $t'(x) \neq \text{constant}, x \in [-1, 1]$.
Based on the feature-centric viewpoint, we will now analyze the properties of $t(x)$. As mentioned in a previous study (Wang et al. 2020), all existing margin-based softmax losses are designed to lower the softmax probability, thereby improving the discriminative power of the learned features. This objective can be formulated as:

$$p_y = \frac{e^{s \cdot \cos\theta_{\boldsymbol{w}_y, \boldsymbol{x}}}}{e^{s \cdot \cos\theta_{\boldsymbol{w}_y, \boldsymbol{x}}} + \sum_{k \neq y}^{K} e^{s \cdot \cos\theta_{\boldsymbol{w}_k, \boldsymbol{x}}}}. \tag{5.11}$$

Based on Eqs. (5.8) and (5.11), we derive the subsequent equation:

$$\begin{aligned} p_y^t &= \frac{e^{s \cdot t\left(\cos\theta_{\boldsymbol{w}_y, \boldsymbol{x}}\right)}}{e^{s \cdot t\left(\cos\theta_{\boldsymbol{w}_y, \boldsymbol{x}}\right)} + e^{s \cdot \cos\theta_{\boldsymbol{w}_y, \boldsymbol{x}}} \left(1 - p_y\right) / p_y} \\ &= \frac{p_y}{p_y + e^{s \cdot \left[\cos\theta_{\boldsymbol{w}_y, \boldsymbol{x}} - t\left(\cos\theta_{\boldsymbol{w}_y, \boldsymbol{x}}\right)\right]} \left(1 - p_y\right)}. \end{aligned} \tag{5.12}$$

To decrease the softmax probability—that is, to ensure $p_y^t \leqslant p_y$—we can derive the following inequalities from Eq. (5.12):

$$\begin{aligned} & p_y^t \leqslant p_y \\ \Longleftrightarrow & p_y + e^{s \cdot \left[\cos\theta_{\boldsymbol{w}_y, \boldsymbol{x}} - t\left(\cos\theta_{\boldsymbol{w}_y, \boldsymbol{x}}\right)\right]} \left(1 - p_y\right) \geqslant 1 \\ \Longleftrightarrow & e^{s \cdot \left[\cos\theta_{\boldsymbol{w}_y, \boldsymbol{x}} - t\left(\cos\theta_{\boldsymbol{w}_y, \boldsymbol{x}}\right)\right]} \geqslant 1 \\ \Longleftrightarrow & \cos\theta_{\boldsymbol{w}_y, \boldsymbol{x}} \geqslant t\left(\cos\theta_{\boldsymbol{w}_y, \boldsymbol{x}}\right). \end{aligned}$$

Based on the preceding analysis, we can now state the third property.

Property 3. $t(x) \leqslant x, x \in [-1, 1]$.
Finally, we analyze the properties of $t(x)$ through an optimization lens. The scale factor s is crucial for margin-based softmax loss because it controls the maximum scale of the similarity score. Specifically, the highest positive similarity value is $s \cdot t(1)$, and the lowest negative similarity value is $-s$. Thus, the range of p_y^t is defined as:

$$\left[\frac{e^{s \cdot t(-1)}}{e^{s \cdot t(-1)} + \sum_{k \neq y}^{K} e^{s}}, \frac{e^{s \cdot t(1)}}{e^{s \cdot t(1)} + \sum_{k \neq y}^{K} e^{-s}}\right].$$

It is evident that when $t(1) \geqslant 0$, the maximum value of p_y^t will be closer to 1 than when $t(1) < 0$. This ensures that when $\cos\theta_{\boldsymbol{w}_y, \boldsymbol{x}} = 1$, the value of Eq. (5.6) will be sufficiently small to allow for more stable convergence. Therefore, it is essential to maintain a large range for p_y^t, similar to the manually designed loss functions shown in Fig. 5.1. From this, we can easily conclude that $t(1) \geqslant 0$, which constitutes our fourth property.

Property 4. $t(1) \geqslant 0$.
By analyzing the margin-based softmax loss function, we have identified four key properties. These same properties must be satisfied by any loss function within our search space. Table 5.1 shows whether the four properties are satisfied by various manually designed loss functions and by AutoML-based loss functions. As the table illustrates, the search space of AM-LFS satisfies only Property 2, which further demonstrates the arbitrary nature of this search space.

In summary, the four essential properties of $t(x)$ are:

- **Property 1:** $t'(x) \geqslant 0, x \in [-1, 1]$.
- **Property 2:** $t'(x) \neq \text{constant}, x \in [-1, 1]$.
- **Property 3:** $t(x) \leqslant x, x \in [-1, 1]$.
- **Property 4:** $t(1) \geqslant 0$.

5.3.3 Sampling Distribution

In this section, we introduce the sampling distribution that will be used within the reinforcement learning framework, as discussed in Sect. 5.3.4. To design a distribution capable of sampling a loss function that satisfies the four properties outlined in Sect. 5.3.2, we model the parameter vector, $\boldsymbol{\Theta} = [a_0, a_1, ..., a_M]^{\mathrm{T}}$, as a set of non-independent truncated Gaussian distributions, which can be expressed as:

$$a_i \sim \begin{cases} \mathcal{TN}(x; \mu_M, \sigma, 0, 1) & , i = M \\ \mathcal{TN}(x; \mu_i, \sigma, -\infty, u_i) & , i < M \end{cases}, \tag{5.13}$$

where $u_i = \min\left(a_{i+1}, \frac{2i}{M} - 1\right)$, and $\mathcal{TN}(x; \mu, \sigma, a, b)$ is the probability density function for the truncated Gaussian distribution (Chopin 2011), which is given by:

$$\mathcal{TN}(x; \mu, \sigma, a, b) = \frac{1}{\sigma} \frac{\phi\left(\frac{x-\mu}{\sigma}\right)}{\Phi(\beta) - \Phi(\alpha)}, \tag{5.14}$$

where $\alpha = (a - \mu)/\sigma$ and $\beta = (b - \mu)/\sigma$ are the normalized bounds, with $-\infty \leqslant a < b \leqslant \infty$ and $a \leqslant x \leqslant b$. The terms μ and σ represent the mean and standard deviation of the underlying normal distribution. The symbols $\phi(x)$ and $\Phi(x)$ refer to the probability density function and cumulative distribution function of the standard normal distribution, respectively:

$$\phi(x) = \frac{\exp\left(-\frac{x^2}{2}\right)}{\sqrt{2\pi}}, \quad \Phi(x) = \frac{\left(1 + \operatorname{erf}\left(\frac{x}{\sqrt{2}}\right)\right)}{2},$$

where erf$(\cdot)$ is the error function. To sample values from the truncated Gaussian distribution, we utilize the inverse transform method (Vogel 2002):

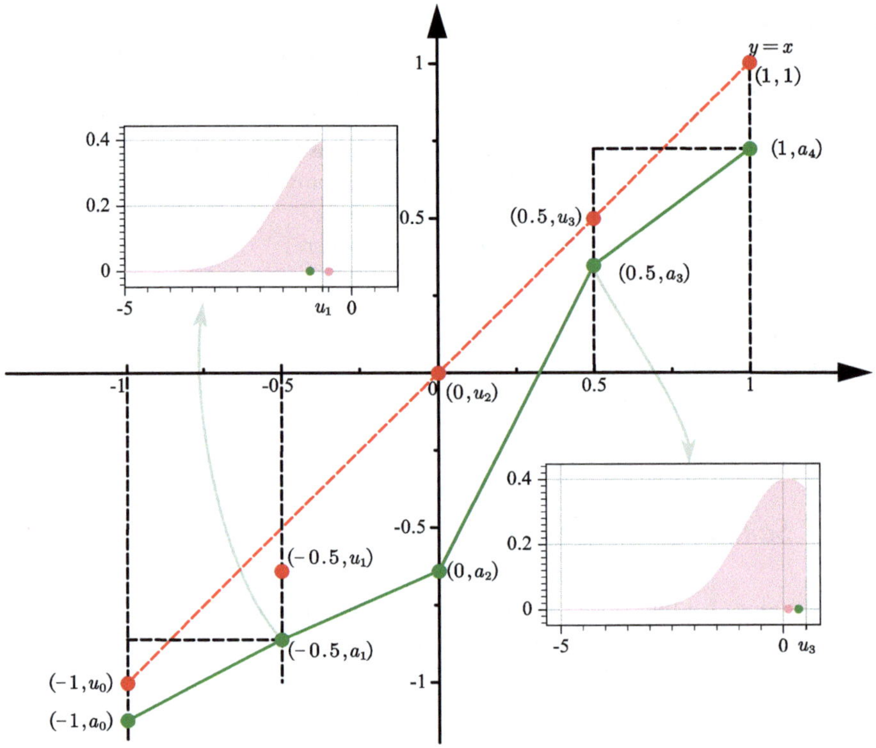

Fig. 5.3 A depiction of the sampling process from $\mathcal{TN}s(\boldsymbol{\mu}, \sigma)$, with $M = 4$. The red dashed line represents $y = x$, while the solid green line shows the resulting sampled $t(x)$. Initially, a_4 is sampled from $\mathcal{TN}(a_4; \mu_4, \sigma, 0, 1)$. Next, $u_3 = \min(a_4, 0.5)$ is computed to determine the sampling distribution $\mathcal{TN}(a_3; \mu_3, \sigma, -\infty, u_3)$. This process is iterated until a_0 is sampled. The smaller insets at the top left and bottom right show the probability density functions for the distributions used to sample a_1 and a_3, respectively. The green dot indicates the sampled point, and the pink dot represents the distribution's mean, μ

$$
\begin{aligned}
x &= g(U) \\
&= \Phi^{-1}\left(\Phi(\alpha) + U \cdot (\Phi(\beta) - \Phi(\alpha))\right)\sigma + \mu,
\end{aligned}
\tag{5.15}
$$

where U is a uniform random variable on $[0, 1]$. By introducing this new random variable U, we can compute the gradients of x with respect to μ, σ, α, and β, making all components differentiable. This approach allows us to establish a relationship among the parameters $a_0, a_1, ..., a_M$ using Eq. (5.13), in contrast to the independent sampling method used in AM-LFS. The specific sampling process is illustrated in Fig. 5.3.

From Eq. (5.15), it is clear that sampling from a truncated Gaussian distribution using the inverse transform method requires the inverse of the erf$(\cdot)$ function. Since there is no closed-form analytical expression for the error function, its inverse is also difficult to compute. As a result, both are commonly approximated using Taylor series, as seen in PyTorch (Paszke et al. 2019). However, this approach can lead to numerical instability, especially when sampling from the tails of the normal distribution (Kroese et al. 2013). To overcome this problem, we propose Theorem 1.

Theorem 1 *Given* $x \sim \mathcal{TN}(x; 0, 1, a, b)$, *with* $-\infty \leqslant a \leqslant x \leqslant b \leqslant \infty$. *If* $0 < a < \infty$ *and* $a \to \infty$, *then* $\mathcal{TN}(x; 0, 1, a, b)$ *approaches the Dirac delta distribution* $\delta(x; a)$. *Similarly, if* $-\infty < b < 0$ *and* $b \to -\infty$, *then* $\mathcal{TN}(x; 0, 1, a, b)$ *also approaches the Dirac delta distribution* $\delta(x; b)$.

The Proof of Theorem 5.1 We first introduce a lemma.

Lemma 1 *According to Eq. (5.14), we can define*

$$\Gamma(x) = \frac{\phi'(x)}{\phi(b) - \phi(a)}, -\infty \leqslant a \leqslant x \leqslant b \leqslant \infty.$$

Then we have

$$\lim_{a \to \infty} \Gamma(x) = \begin{cases} \infty, & x = a \\ 0, & x > a \end{cases}.$$

Proof By definition,

$$\begin{aligned} \Gamma(x) &= \frac{\phi'(x)}{\phi(b) - \phi(a)} \\ &= \frac{xe^{-\frac{x^2}{2}}}{e^{-\frac{a^2}{2}} - e^{-\frac{b^2}{2}}}. \end{aligned}$$

When $x = a$, we evaluate the limit as $a \to \infty$:

$$\begin{aligned} \lim_{a \to \infty} \Gamma(a) &= \lim_{a \to \infty} \frac{ae^{-\frac{a^2}{2}}}{e^{-\frac{a^2}{2}} - e^{-\frac{b^2}{2}}} \\ &= \lim_{a \to \infty} \frac{a}{1 - e^{\frac{a^2 - b^2}{2}}} \\ &> \lim_{a \to \infty} a = \infty. \end{aligned}$$

When $x > a$, that is, $x = a + \epsilon, \epsilon > 0$, then we have

$$\begin{aligned} \lim_{a \to \infty} \Gamma(a + \epsilon) &= \lim_{a \to \infty} \frac{(a + \epsilon) e^{-\frac{(a+\epsilon)^2}{2}}}{e^{-\frac{a^2}{2}} - e^{-\frac{b^2}{2}}} \\ &= \lim_{a \to \infty} \frac{a + \epsilon}{e^{\frac{\epsilon^2 + 2\epsilon a}{2}} - e^{\frac{(a+\epsilon-b)(a+\epsilon+b)}{2}}} \\ &\leqslant \lim_{a \to \infty} \frac{a + \epsilon}{e^{\frac{\epsilon^2 + 2\epsilon a}{2}} - 1}. \end{aligned}$$

$\forall \epsilon > 0$, $\lim_{a\to\infty} a + \epsilon \to \infty$ and $\lim_{a\to\infty} e^{\frac{\epsilon^2+2\epsilon a}{2}} - 1 \to \infty$, L'Hôpital's rule (Boyer and Merzbach 2011) can be used to get

$$\begin{aligned}\lim_{a\to\infty} \Gamma(a+\epsilon) &\leqslant \lim_{a\to\infty} \frac{a+\epsilon}{e^{\frac{\epsilon^2+2\epsilon a}{2}} - 1} \\ &= \lim_{a\to\infty} \frac{1}{2\epsilon e^{\frac{\epsilon^2+2\epsilon a}{2}}} = 0.\end{aligned}$$

Thus, we conclude that

$$\lim_{a\to\infty} \Gamma(x) = \begin{cases} \infty & , x = a \\ 0 & , x > a \end{cases}.$$

□

Now we begin to proof the Theorem 1.

Proof Here we focus on the case where $a \to \infty$; the case for $b \to -\infty$ can be demonstrated analogously. From Eq. (5.14), we have

$$\mathcal{TN}(x; 0, 1, a, b) = \frac{\phi(x)}{\Phi(b) - \Phi(a)},$$

then we define

$$\tilde{\delta}(x; a) = \begin{cases} 0 & , x < a \\ \frac{\phi(x)}{\Phi(b)-\Phi(a)} & , x \geqslant a \end{cases},$$

Since $x \geqslant a$, and because $\lim_{a\to\infty} \phi(x) \to 0$ and $\lim_{a\to\infty} \Phi(b) - \Phi(a) \to 0$, L'Hôpital's rule (Boyer and Merzbach 2011) can also be applied to derive the following:

$$\lim_{a\to\infty} \tilde{\delta}(x; a) = \lim_{a\to\infty} \frac{\phi'(x)}{\phi(b) - \phi(a)}, x \geqslant a$$

According to Lemma 1, we have

$$\lim_{a\to\infty} \tilde{\delta}(x; a) = \begin{cases} \infty & , x = a \\ 0 & , x \neq a \end{cases}. \tag{5.16}$$

According to the definition of $\tilde{\delta}(x; a)$, we have $\forall a$, $\int_{-\infty}^{\infty} \tilde{\delta}(x; a) = 1$, so

$$\int_{-\infty}^{\infty} \lim_{a\to\infty} \tilde{\delta}(x; a) = 1. \tag{5.17}$$

Equations (5.16) and (5.17) together satisfy the definition of the Dirac delta distribution. Consequently, as $a \to \infty$, $\tilde{\delta}(x; a)$ converges to the Dirac delta distribution $\delta(x; a)$. □

According to Theorem 1, we can modify Eq. (5.15) as follows:

$$x = \begin{cases} \alpha\sigma + \mu & , \alpha \geqslant \tilde{\alpha} \\ \beta\sigma + \mu & , \beta \leqslant \tilde{\beta} \\ g(U) & , \text{otherwise} \end{cases}, \tag{5.18}$$

where $\tilde{\alpha} > 0$ and $\tilde{\beta} < 0$ are predefined constants. When $\alpha \geqslant \tilde{\alpha}$, the effective range of the truncated distribution is located in the far right tail of the standard normal distribution. As α increases, the truncated Gaussian distribution can be approximated by a Dirac delta distribution. A similar approximation is valid when $\beta \leqslant \tilde{\beta}$. Theoretically, the absolute values of $\tilde{\alpha}$ and $\tilde{\beta}$ should be as large as possible. However, setting them too high would not address the numerical instability problem during sampling. In this work, we determined that an absolute value of 5 for both $\tilde{\alpha}$ and $\tilde{\beta}$ effectively resolves this instability.

For convenience, the distribution used for sampling $\mathbf{\Theta} = [a_0, a_1, ..., a_M]^{\mathrm{T}}$ is denoted as

$$\mathbf{\Theta} \sim \mathcal{TN}s(\boldsymbol{\mu}, \sigma), \tag{5.19}$$

where $\boldsymbol{\mu} = [\mu_0, \mu_1, ..., \mu_M]$.

In Table 5.1, we compare three AutoML-based loss functions. Regarding the expressiveness of $t(x)$, AM-LFS possesses the most diverse search space, followed by our method, while SS is the most constrained. The extremely limited expressive power of the SS method's search space severely hampers its performance, notwithstanding its high stability. Although our search space is less diverse than that of AM-LFS, our approach adheres to four critical properties, ensuring that all sampled loss functions are effective. As a result, our search space achieves a balance between high stability and a sufficient level of diversity.

5.3.4 Optimization

Suppose we have a network model $\mathcal{M}_{\mathbf{\Omega}}$ with parameters $\mathbf{\Omega}$, and training and validation datasets $\mathcal{D}_{trn}$ and $\mathcal{D}_{val}$, respectively. The objective of loss function search is to maximize the rewards $r(\mathcal{M}_{\mathbf{\Omega}}; \mathcal{D}_{val})$ of the model $\mathcal{M}_{\mathbf{\Omega}}$ on the validation dataset $\mathcal{D}_{val}$, with respect to the parameters $\mathbf{\Theta} = [a_0, a_1, ..., a_M]^{\mathrm{T}}$. The model $\mathcal{M}_{\mathbf{\Omega}}$ is acquired by minimizing the loss function $\mathcal{L}^{\mathbf{\Theta}}$:

$$\begin{aligned} \mathbf{\Theta}^* &= \arg\max_{\mathbf{\Theta}} r\left(\mathcal{M}_{\mathbf{\Omega}^*(\mathbf{\Theta})}; \mathcal{D}_{val}\right) \\ \text{s.t. } \mathbf{\Omega}^*(\mathbf{\Theta}) &= \arg\min_{\mathbf{\Omega}} \sum_{(\boldsymbol{x},y)\in\mathcal{D}_{trn}} \mathcal{L}^{\mathbf{\Theta}}\left(\mathcal{M}_{\mathbf{\Omega}}(\boldsymbol{x}), y\right). \end{aligned} \tag{5.20}$$

The optimization challenge outlined above is a classic bi-level optimization problem (Colson et al. 2007). The inner optimization loop focuses on determining the

Algorithm 1 LFS-ReID

Require: Training dataset $\mathcal{D}_{trn}$, validation dataset $\mathcal{D}_{val}$, initialized model $\mathcal{M}_{\boldsymbol{\Omega}_0}$, initialized distribution $\mathcal{TN}s(\boldsymbol{\mu}_0, \sigma)$, total training epochs T, distribution learning rate τ, sampling number B.
Ensure: Final model $\mathcal{M}_{\boldsymbol{\Omega}_E}$.

```
i ← 0
M'_{Ω_0^i} ← M_{Ω_0}
for t = 1 to T do
    for b = 1 to B do
        Update M_{Ω_{t-1}^b} with M'_{Ω_{t-1}^i}
    end for
    Shuffle the training dataset D_trn
    for b = 1 to B (synchronously in parallel) do
        Sample Θ^b from TNs(μ_{t-1}, σ)
        Train the model M_{Ω_{t-1}^b} with Θ^b for one epoch
        Get trained model M'_{Ω_t^b}
        Calculate the reward r(M'_{Ω_t^b(Θ^b)}; D_val)
    end for
    Update μ_t using Eq. (5.21)
    i ← arg max_{b∈[1,B]} r(M'_{Ω_t^b(Θ^b)}; D_val)
end for
return M_{Ω_E} = M'_{Ω_T^i}
```

optimal model parameters, $\boldsymbol{\Omega}^*(\boldsymbol{\Theta})$, by minimizing the training loss $\mathcal{L}^{\boldsymbol{\Theta}}$ for a specific set of hyperparameters, $\boldsymbol{\Theta}$. The outer optimization loop then identifies the optimal hyperparameters, $\boldsymbol{\Theta}^*$, by maximizing the reward $r\left(\mathcal{M}_{\boldsymbol{\Omega}^*(\boldsymbol{\Theta})}; \mathcal{D}_{val}\right)$ on the validation dataset. For this chapter, the mAP score on the validation dataset serves as our reward function $r\left(\mathcal{M}_{\boldsymbol{\Omega}^*(\boldsymbol{\Theta})}; \mathcal{D}_{val}\right)$.

Instead of using a single, fixed loss function, we adopt an online hyperparameter optimization method similar to AM-LFS (Li et al. 2019). Specifically, in each training epoch, we sample B sets of hyperparameters, $\boldsymbol{\Theta}^1$, $\boldsymbol{\Theta}^2$, ..., $\boldsymbol{\Theta}^B$, from $\mathcal{TN}s(\boldsymbol{\mu}, \sigma)$. We then train B models with identical initial parameters for one epoch in parallel, each using a different set of hyperparameters.[2]

After one epoch, we obtain B rewards on the validation dataset from the B trained models. Using these rewards, we update the parameters of the sampling distribution, $\mathcal{TN}s(\boldsymbol{\mu}, \sigma)$, via REINFORCE (Williams 1992) as follows:

$$\boldsymbol{\mu}_t = \boldsymbol{\mu}_{t-1} + \tau \frac{1}{B} \sum_{b=1}^{B} \tilde{r}(\boldsymbol{\Theta}^b) \nabla_{\boldsymbol{\Theta}} \log p(\boldsymbol{\Theta}^b), \tag{5.21}$$

[2] To eliminate randomness from data loading, we ensure that the data fed to all B models is consistent.

Table 5.2 Dataset statistics. T: Train. Q: Query. G: Gallery

Dataset	# IDs (T-Q-G)	# images (T-Q-G)
Market-1501	751-750-751	12936-3368-15913
CUHK03	767-700-700	7365-1400-5332
MSMT17	1041-3060-3060	32621-11659-82161

where $p(\boldsymbol{\Theta}^b)$ is the PDF of $\mathcal{TN}s(\boldsymbol{\Theta}; \boldsymbol{\mu}_{t-1}, \sigma)$. $\tilde{r}(\boldsymbol{\Theta}^b)$ is the normalized reward.[3] The model with the highest reward is used to initialize the B models for the next epoch.

Once training has converged, we select the model with the highest score $r\left(\mathcal{M}_{\boldsymbol{\Omega}^*(\boldsymbol{\Theta})}; \mathcal{D}_{val}\right)$ as our final model, without any need for re-training. To simplify the problem, we keep σ fixed as a constant and only optimize $\boldsymbol{\mu}$. For clarity, the training procedure for our LFS-ReID method is summarized in Algorithm 1.

5.4 Experiments

5.4.1 Datasets and Evaluation Metrics

Datasets. To validate the performance of our model, we conducted experiments on three widely used datasets (see Table 5.2 for statistics):

Market-1501 (Zheng et al. 2015) is a large-scale dataset for person ReID. It comprises 19,372 gallery images, 3,368 query images, and 12,396 training images captured by six cameras. The dataset contains 751 training identities and 750 test identities, with no overlap between the two sets. On average, each training identity has 17.2 images.

CUHK03 (Li et al. 2014) contains 1,467 unique identities and 28,192 bounding boxes. The dataset is divided into a training set with 26,264 images from 1,367 identities and a test set with 1,928 images from 100 identities. We adopt the new data split protocol from Zhong et al. (2017). CUHK03 provides two versions: one with manual labels and one with automatic detections. We use the more challenging automatically detected version for our experiments, as it better reflects real-world application scenarios.

MSMT17 (Wei et al. 2018) is currently the largest person ReID dataset, featuring 126,441 images of 4,101 identities from 15 cameras. The dataset is partitioned into a training set with 32,621 bounding boxes of 1,041 identities and a test set with 93,820 bounding boxes of 3,060 identities. For the test set, 11,659 images are designated as queries, and the remaining 82,161 are used as gallery images. This dataset is particularly challenging due to its more complex and diverse scenes, which include both indoor and outdoor environments.

[3] We normalized the rewards to have a zero mean.

Evaluation Metrics. To assess the performance of our LFS-ReID method and compare it with other ReID approaches, we report two standard evaluation metrics: the cumulative matching characteristic (CMC) at rank-1 and mean average precision (mAP). We follow the common evaluation settings for all three benchmarks (Zheng et al. 2015; Wei et al. 2018).

5.4.2 Implementation Details

Since our method is orthogonal to the specific backbone network architecture, we adopt four common models to comprehensively verify its effectiveness: ResNet50 (He et al. 2016), OSNet (Zhou et al. 2019), MGN (Wang et al. 2018a), and Pyramid (Zheng et al. 2019b). The first two models are based on global features, while the latter two are based on local features. Specifically, the feature dimension of OSNet is 512. For ResNet50, we obtain a 2048-dimension feature by adding a batch normalization layer (Ioffe and Szegedy 2015) after the average pooling. For MGN and Pyramid, we apply LFS-ReID only to the softmax loss branches, keeping the triplet loss unchanged, similar to the approach in AM-LFS. The feature dimensions for MGN and Pyramid are 2048 and 2688, respectively.

- **LFS-ReID hyperparameters.** We set σ^2 to 0.2 and use the Adam optimizer with $\tau = 0.05$ to update the distribution parameters, consistent with AM-LFS. The values of $\tilde{\alpha}$ and $\tilde{\beta}$ are set to 5 and -5, respectively. Unless otherwise specified, we use $M = 6$, $B = 8$, and $s = 64.0$.
- **Data preprocessing.** For ResNet50 and OSNet, we resize each image to 256×128. For MGN and Pyramid, we follow their original papers and use an image size of 384×128. Data augmentation techniques such as random cropping, random horizontal flipping, and random erasing (Zhong et al. 2020) are also applied.
- **Data sampling.** We use the PK sampling strategy (Fan et al. 2019) with $P = 16$ and $K = 4$.
- **Training.** We use the SGD optimizer with a momentum of 0.9 and a weight decay of 5e-4 to train all four models. All models are trained for a total of 140 epochs, employing a warmup learning rate strategy (Luo et al. 2020). We linearly increase the learning rate from 0.0001 to 0.01 over the first 20 epochs, during which the backbone network weights are frozen. The learning rate is then decayed to 0.001 and 0.0001 at the 60th and 100th epochs, respectively.
- **Testing.** During testing, we utilize test-time flipping and cosine similarity as the measurement metric (Luo et al. 2020). For all datasets, each sampled model is trained on one TITAN X GPU, for a total of 8 GPUs. All experiments in this chapter are implemented using PyTorch (Paszke et al. 2019). We use Ray (Moritz et al. 2018) to implement LFS-ReID, which allows for distributed model training.

5.4.3 Comparison with State-of-the-Art Methods

Our comparison with state-of-the-art methods is categorized into two types: models that use global features and models that use local features. Most global feature-based models utilize ResNet50 as their backbone. We also include architectures based on manual design, such as OSNet (Zhou et al. 2019), and AutoML-based architectures like AutoReID (Quan et al. 2019). For local feature-based models, our primary comparison is with methods using MGN (Wang et al. 2018a) as the backbone, and we also include other local feature models like PCB (Sun et al. 2018), Pyramid (Zheng et al. 2019b), and ABD-Net (Chen et al. 2019). The comparison results are presented in Table 5.3.

Among the global-feature-based models, our method achieves the highest mAP on all three datasets. Specifically, on the Market-1501 dataset, our method combined with ResNet50 yields a mAP 0.7% higher than the second-best, OSNet (Zhou et al. 2019). When our method is integrated with OSNet, it improves the mAP by 2.2% compared to the original OSNet. On the largest dataset, MSMT17, our method surpasses the second-best OSNet by 1.3% and 3.2% when using ResNet50 and OSNet, respectively. It should be noted that OSNet and AutoReID (Quan et al. 2019) are based on manually and AutoML-designed network architectures, respectively, which generally outperform ResNet50. As the results in Table 5.3 demonstrate, our method's performance with OSNet is indeed significantly better than with ResNet50 across all three datasets.

In comparison with local-feature-based models, our method, when combined with MGN, achieves the top ranking on the Market-1501 dataset, with an mAP of 89.4% and a rank-1 accuracy of 95.8%. On the CUHK03 dataset, our method combined with MGN improves mAP by 7.6% over the original MGN, and our method with Pyramid improves mAP by 3.0% compared to the original Pyramid. On the largest MSMT17 dataset, our method also ranks first, achieving 62.3% mAP and 63.5% rank-1 accuracy when using MGN and Pyramid, respectively.

5.4.4 Ablation Study

Comparison with variants of softmax loss. To further demonstrate the effectiveness of our method, we conducted a comparison between the baseline softmax loss, several hand-designed methods (including Sphere-Softmax (Fan et al. 2019), AM-Softmax (Wang et al. 2018c, b), Arc-Softmax (Deng et al. 2019), and CircleLoss (Sun et al. 2020)), and two AutoML-based methods (AM-LFS (Li et al. 2019) and SS (Wang et al. 2020)). The comparisons were performed using the same model and training settings. We were able to download the source code for the hand-designed methods from their respective authors' or GitHub pages. However, since no public code is available for AM-LFS and SS, we did our best to re-implement them.

Table 5.3 A comparison with state-of-the-art methods. The models above the double line are based on global features, while those below are based on local features. Within each section, our methods are listed below the single line, and other methods are above. The number in parentheses following the loss indicates its usage count. **Red bold** highlights our best performance, and **Blue bold** indicates the best performance among other methods

Methods	Loss	Market-1501		CUHK03		MSMT17	
		mAP	Rank-1	mAP	Rank-1	mAP	Rank-1
TriNet (Hermans et al. 2017)	Triplet	69.1	84.9	–	–	–	–
SphereReID‡ (Fan et al. 2019)	Sphere-Softmax	83.6	94.4	–	–	–	–
StrongBaseline (Luo et al. 2019)	Softmax+Triplet	85.9	94.5	–	–	–	–
ResNet50+CircleLoss (Sun et al. 2020)	CircleLoss	84.9	94.2	–	–	50.2	76.3
OSNet (Zhou et al. 2019)	Softmax	**86.7**	94.8	67.8	72.3	**55.1**	**79.1**
AutoReID (Quan et al. 2019)	Softmax+Triplet	85.1	94.5	**69.3**	**73.3**	52.5	78.2
SphereReID+AM-LFS‡ (Li et al. 2019)	AM-LFS	84.4	**95.0**	–	–	–	–
ResNet50+LFS-ReID(Ours)	LFS-ReID	87.4	95.2	72.3	75.4	56.4	79.9
OSNet+LFS-ReID(Ours)	LFS-ReID	**88.9**	**95.7**	**75.4**	**78.8**	**58.3**	**81.7**
PCB+RPP (Sun et al. 2018)	Softmax(6)	81.6	93.8	57.5	63.7	–	–
MGN (Wang et al. 2018a)	Softmax(8)+Triplet(3)	86.9	95.7	66.0	66.8	–	–
MGN+CircleLoss (Sun et al. 2020)	CircleLoss	87.4	**96.1**	–	–	52.1	76.9
Pyramid (Zheng et al. 2019b)	Softmax(21)+Triplet	88.2	95.7	**74.8**	**78.9**	–	–
ABD-Net (Chen et al. 2019)	Softmax(2)	**88.3**	95.6	–	–	**60.8**	**82.3**
MGN+AM-LFS‡ (Li et al. 2019)	AM-LFS(8)+Triplet(3)	88.1	95.8	–	–	–	–
MGN+AM-LFS(Our Imp.)	AM-LFS(8)+Triplet(3)	88.3	95.4	72.0	75.2	59.2	80.9
MGN+LFS-ReID(Ours)	LFS-ReID(8)+Triplet(3)	**89.4**	**95.8**	73.6	76.9	62.3	82.6
Pyramid+LFS-ReID(Ours)	LFS-ReID(21)+Triplet	88.7	95.6	**77.8**	**80.0**	**63.5**	**83.3**

‡ The official code has not been released. We report the numbers directly from the original paper, as they surpass the results of our re-implementation

Table 5.4 Comparison among various softmax loss functions using the ResNet50 baseline model

Methods	Market-1501		CUHK03	
	mAP	Rank-1	mAP	Rank-1
Softmax	70.95	87.50	28.31	27.07
Sphere-Softmax	82.01	93.11	55.59	57.50
AM-Softmax	84.41	94.24	49.47	52.07
Arc-Softmax	84.80	94.54	59.98	62.29
CircleLoss	85.12	94.33	62.69	65.57
AM-LFS	86.53	94.53	67.88	70.29
SS	85.68	94.39	63.11	65.79
Ours	**87.39**	**95.19**	**72.27**	**75.43**

We used a grid search method to determine the hyperparameters for each loss function. The selected hyperparameters for each loss are as follows: $s = 16$ for Sphere-Softmax, $s = 128, m = 0.5$ for AM-Softmax, $s = 128, m = 0.45$ for Arc-Softmax, and $s = 128, m = 0.25$ for CircleLoss. The scale factor s for AM-LFS, SS, and our method is set to 64. A comparison between our method and the other variants of softmax loss, under identical model and training settings, is presented in Table 5.4.

From Table 5.4, it can be concluded that the results obtained with the standard softmax loss are inferior to those of its variants, which highlights that features learned with margin-based softmax loss functions are indeed more discriminative. Furthermore, the AutoML-based methods outperform the hand-designed methods. Among the three AutoML-based approaches, our method achieves the highest mAP and Rank-1 accuracy on both the Market-1501 and CUHK03 datasets, confirming that our approach is more effective.

Investigation on the stability. We compare the stability of each method under different initializations. For each experiment, we ensure that the network weight initialization and data loading order are consistent. A total of five experiments are conducted for each method, with all settings being identical to those previously described, except for the initial state.

From the average performance of each method presented in Table 5.5, our method performed the best, followed by SS, with AM-LFS showing the worst performance. When judging by the best performance out of the five experiments, our method remains superior, followed by AM-LFS, and SS has the lowest peak performance.

In terms of stability, SS exhibits the highest stability, followed by our method, while AM-LFS is the least stable. Although our method's stability is slightly weaker than that of SS, it is a dramatic improvement over AM-LFS. Most importantly, the overall performance of SS is clearly inferior to our method. This is because the diversity of SS's search space is too limited, whereas our search space has a strong ability to express $t(x)$. This once again validates that our search space offers the dual advantages of higher diversity and superior stability.

Table 5.5 Analysis of stability in mAP and rank-1 on the Market-1501 and CUHK03 datasets using the ResNet50 baseline model. Five separate experiments were conducted for each method, with identical settings as previous tests, except for the initialization. **Bold** denotes the best performance among the three methods (AM-LFS, SS, and ours). "mean" indicates the average performance across the five experiments. "std" represents the standard deviation of the performance, and "best" shows the highest performance achieved

Dataset	Experiments	AM-LFS		SS		Ours	
		mAP	Rank-1	mAP	Rank-1	mAP	Rank-1
Market-1501	#1	86.53	94.53	85.68	94.39	87.39	95.19
	#2	85.58	94.51	85.44	94.26	87.22	95.00
	#3	74.52	89.04	85.65	94.56	87.24	94.89
	#4	75.96	88.90	85.23	94.36	86.59	94.21
	#5	85.55	94.15	85.30	94.40	86.86	94.77
	Mean	81.63	92.23	85.46	94.39	**87.06**	**94.81**
	Std	5.87	2.98	**0.20**	**0.11**	0.33	0.37
	Best	86.53	94.53	85.68	94.56	**87.39**	**95.19**
CUHK03	#1	59.87	62.07	62.95	65.36	71.00	73.71
	#2	50.27	51.50	63.05	65.07	70.95	74.14
	#3	57.80	59.57	63.11	65.21	71.39	73.79
	#4	67.88	70.29	63.11	65.79	72.27	75.43
	#5	65.13	67.21	62.98	65.00	70.53	72.50
	Mean	60.19	62.13	63.04	65.29	**71.23**	**73.91**
	Std	6.85	7.28	**0.07**	**0.31**	0.66	1.05
	Best	67.88	70.29	63.11	65.79	**72.27**	**75.43**

Effectiveness of the properties in $t(x)$. We compared the four properties identified in Sect. 5.3.2. The specific sampling distribution for each setting is detailed in Table 5.6. Four key observations can be made from Table 5.7:

- When the search space is completely unrestricted, as in setting 1, the result is the worst.
- When Property 2 is not considered, as in settings 2, 3, 4, and 5, there remains a noticeable performance gap compared to cases where Property 2 is included, such as settings 6, 7, 8, and 9.
- While Property 3 yields a more significant improvement on its own than Property 4, the greatest benefit is achieved when both are considered together.
- The best result is obtained when all four properties are taken into account. Therefore, we can conclude that all four of these properties play a crucial role.

Investigation on the number of sampled models. We studied the effect of the number of samples in the optimization procedure by varying the parameter B in AM-LFS, SS, and our method. As shown in Table 5.8, the mAP of all three methods on the Market-1501 dataset, using the ResNet50 model, improves as the number of

Table 5.6 An analysis of how sampling distributions influence the effectiveness of properties. "P1", "P2", "P3", and "P4" denote "Property 1", "Property 2", "Property 3", and "Property 4", respectively. The variable u_i is defined as $u_i = \min\left(a_{i+1}, \frac{2i}{M} - 1\right)$

Setting	P1	P2	P3	P4	$t(x)$	Sampling distribution
1	✗	✗	✗	✗	$pwl(x; a_0, a_1)$	$a_1 \sim \mathcal{N}(\mu_1, \sigma^2), a_0 \sim \mathcal{N}(\mu_0, \sigma^2)$
2	✓	✗	✗	✗	$pwl(x; a_0, a_1)$	$a_1 \sim \mathcal{N}(\mu_1, \sigma^2), a_0 \sim \mathcal{TN}(\mu_0, \sigma, -\infty, a_1)$
3	✓	✗	✓	✗	$pwl(x; a_0, a_1)$	$a_1 \sim \mathcal{TN}(\mu_1, \sigma, -\infty, 1), a_0 \sim \mathcal{TN}(\mu_0, \sigma, -\infty, u_0)$
4	✓	✗	✗	✓	$pwl(x; a_0, a_1)$	$a_1 \sim \mathcal{TN}(\mu_1, \sigma, 0, \infty), a_0 \sim \mathcal{TN}(\mu_0, \sigma, -\infty, a_1)$
5	✓	✗	✓	✓	$pwl(x; a_0, a_1)$	$a_1 \sim \mathcal{TN}(\mu_1, \sigma, 0, 1), a_0 \sim \mathcal{TN}(\mu_0, \sigma, -\infty, u_0)$
6	✓	✓	✗	✗	$pwl(x; a_0, a_1, ...a_6)$	$a_6 \sim \mathcal{N}(\mu_6, \sigma^2), a_i \sim \mathcal{TN}(\mu_i, \sigma, -\infty, a_{i+1}), i = 5, 4, 3, 2, 1, 0$
7	✓	✓	✓	✗	$pwl(x; a_0, a_1, ...a_6)$	$a_6 \sim \mathcal{TN}(\mu_6, \sigma, -\infty, 1), a_i \sim \mathcal{TN}(\mu_i, \sigma, -\infty, u_i), i = 5, 4, 3, 2, 1, 0$
8	✓	✓	✗	✓	$pwl(x; a_0, a_1, ...a_6)$	$a_6 \sim \mathcal{TN}(\mu_6, \sigma, 0, \infty), a_i \sim \mathcal{TN}(\mu_i, \sigma, -\infty, a_{i+1}), i = 5, 4, 3, 2, 1, 0$
9	✓	✓	✓	✓	$pwl(x; a_0, a_1, ...a_6)$	$a_6 \sim \mathcal{TN}(\mu_6, \sigma, 0, 1), a_i \sim \mathcal{TN}(\mu_i, \sigma, -\infty, u_i), i = 5, 4, 3, 2, 1, 0$

Table 5.7 Effects of the four properties on mAP and Rank-1 performance on the Market-1501 dataset, using the ResNet50 baseline model

Setting	P1	P2	P3	P4	mAP	Rank-1
1	✗	✗	✗	✗	85.48	94.11
2	✓	✗	✗	✗	85.59	94.26
3	✓	✗	✓	✗	85.76	94.29
4	✓	✗	✗	✓	85.68	94.21
5	✓	✗	✓	✓	85.97	94.32
6	✓	✓	✗	✗	86.33	94.34
7	✓	✓	✓	✗	86.85	94.87
8	✓	✓	✗	✓	86.59	94.53
9	✓	✓	✓	✓	**87.39**	**95.19**

Table 5.8 Impact of the number of samples, with B set to 2, 4, 8, and 16, on mAP performance for three AutoML-based methods on the Market-1501 dataset, using the ResNet50 baseline model

Methods	$B = 2$	$B = 4$	$B = 8$	$B = 16$
AM-LFS	78.56	84.15	86.53	86.46
SS	84.53	85.41	85.68	85.81
Ours	86.04	87.03	87.39	87.33

Table 5.9 The impact of the number of intervals, M, on mAP performance on the Market-1501 dataset. The analysis uses three AutoML-based methods with a ResNet50 baseline model, with M set to 1, 3, 6, and 10

Methods	$M = 1$	$M = 3$	$M = 6$	$M = 10$
AM-LFS	85.54	86.66	86.53	86.40
SS	85.68	–	–	–
Ours	85.97	86.85	87.39	86.66

sampled models increases. When the value of B is 16, the performance of AM-LFS and our method has already saturated, with SS showing only a 0.13% improvement compared to when $B = 8$. To achieve a good balance between performance and training efficiency, we chose to fix B at eight for our training.

Investigation on the number of intervals. Because our approach, like AM-LFS, uses a piecewise linear function for $t(x)$, the number of piecewise intervals, M, is a critical parameter that needs to be properly set. Table 5.9 presents the mAP performance results on the Market-1501 dataset with varying values of M. When $M = 1$, both AM-LFS and our method are equivalent to SS and do not satisfy Property 2. In this specific scenario, our method still achieves the highest mAP among the three methods, though the difference is minimal. As M increases, the mAP for both our method and AM-LFS first improves and then starts to decrease. A small M (e.g., $M = 3$) results in a low mAP because the number of intervals is insufficient to address all the hard examples encountered during training. Conversely, a large M (e.g., $M = 10$) leads to an excessive number of distribution parameters to optimize, which also contributes to a lower mAP. We found that the optimal performance is achieved at a moderate value of $M = 6$.

Impact of the hyperparameters. We begin by examining the impact of the scale factor s from Eq. (5.2). The scale factor s is crucial in margin-based softmax loss as it governs the maximum magnitude of the similarity score between $\boldsymbol{w}_y$ and $\boldsymbol{x}$. We experimentally assessed its effect on LFS-ReID and compared it with Sphere-Softmax, AM-Softmax, Arc-Softmax, CircleLoss, AM-LFS, and SS. For AM-Softmax, Arc-Softmax, and CircleLoss, which have an extra hyperparameter m, we report the result corresponding to the optimal m for the given s. We evaluated the effect of s on each loss function at values of 8, 16, 32, 64, 128, and 256, as illustrated in Fig. 5.4a. Our method exhibits the highest robustness for s values ranging from 8 to 128. At larger values of s, our method's performance is slightly lower than that of CircleLoss. The

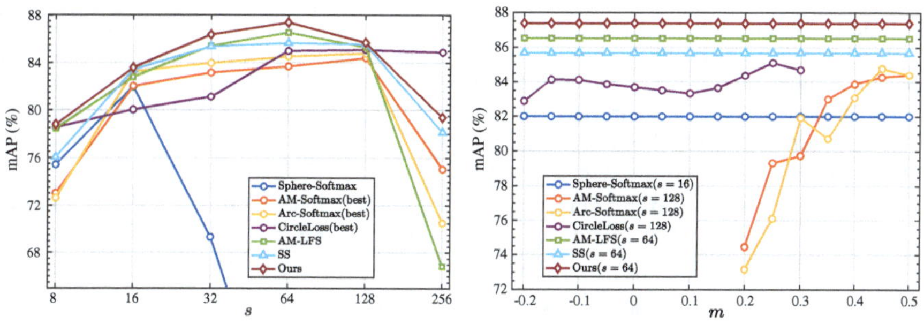

(a) Impact of the s

(b) Impact of the m

Fig. 5.4 The effect of hyperparameters on the mAP performance of various softmax loss function variants on the Market-1501 dataset, using the ResNet50 baseline model

Table 5.10 Training costs (TITAN X GPU) on the Market-1501 dataset with the ResNet50 baseline model

Methods	Wall-clock Time	GPU usage	Total time
Sphere-Softmax	2.74 h	1	2.74×6 h
CircleLoss	2.62 h	1	$2.62 \times 6 \times 11$ h
Ours	3.17 h	8	3.17×6 h

performance gap between CircleLoss and other margin-based softmax losses persists because CircleLoss also optimizes the negative similarity function. Overall, within the framework of margin-based softmax loss functions, our method demonstrates the greatest robustness across various settings of the scale factor s. Our method, similar to AM-LFS, SS, and Sphere-Softmax, does not require tuning an additional hyperparameter m. As shown in Fig. 5.4b, the AutoML-based methods significantly outperform the hand-designed ones.

Analysis of the training cost. We compared the training time of our method with hand-designed methods, as shown in Table 5.10. The training times for AutoML methods are similar, so we only present the results for our method. The same applies to AM-Softmax, Arc-Softmax, and CircleLoss, so we only included CircleLoss in our comparison. Based on the wall-clock time in Table 5.10, our method requires more training time. The main reason for this is the need to calculate mAP on the validation dataset during each epoch. If this mAP calculation time is excluded, our training time is comparable to that of the other two methods. In terms of hardware, our method utilizes 8 GPUs, while the hand-designed loss functions only require one. However, when considering the total time spent on hyperparameter tuning for each method (refer to Fig. 5.4), Sphere-Softmax and our method are more efficient. They only require tuning a single hyperparameter, s (from 8, 16, 32, 64, 128, 256). As a result, the total time needed is approximately one-tenth of the time required for CircleLoss, which must tune two hyperparameters: s (from 8, 16, 32, 64, 128, 256) and m (from –0.2 to 0.3 in increments of 0.05). Even if we account for the 8

Table 5.11 The effectiveness of the scheduled sampling distribution in terms of mAP and rank-1 on CUHK03, using the ResNet50 baseline model. "Scheduled" denotes the sampling distribution with dynamic optimization, which is our LFS-ReID method. "Fixed-N" refers to a distribution where the parameters are fixed at their values from the N-th epoch of the "Scheduled" optimization. The training procedure for "Fixed-N" is identical to "Scheduled," with the sole exception that the sampling distribution parameters are held constant

	Scheduled	Fixed-0	Fixed-20	Fixed-40	Fixed-60	Fixed-80	Fixed-100	Fixed-120	Fixed-140
mAP	72.27	69.62	70.86	70.42	71.31	70.57	71.42	70.58	70.79
Rank-1	75.43	73.00	74.00	74.07	74.00	73.43	75.00	73.14	73.57

GPUs used by our method, the total time is still more efficient than the time spent on CircleLoss. Compared to Sphere-Softmax, our method can increase the mAP on the Market-1501 dataset by 5.38% (as shown in Table 5.4), which we believe justifies the additional computational cost.

Effectiveness of the scheduled sampling distribution. We compared the scheduled sampling distribution with the fixed sampling distribution, as shown in Table 5.11. For all experiments, we ensured consistent network weight initialization and data loading order. As the table indicates, the scheduled method achieves superior performance, demonstrating that dynamically adjusting the sampling distribution's parameters throughout training is more effective. We also found that even with fixed sampling distribution parameters, good performance can still be attained, which further confirms the effectiveness of our search space.

Investigation on the retrieval results. We compared the retrieval results of AM-LFS, SS, and our method using the MGN model on the Market-1501 dataset. As shown in Fig. 5.5, panels (a) and (b) represent relatively simple query images, whereas (c) and (d) depict more challenging scenarios. In the simpler cases, it is clear that the retrieval results from the SS method are slightly worse than those of AM-LFS and our method. In the more difficult situations presented in (c) and (d), our method consistently achieves the best retrieval results.

Nevertheless, because no extra annotation information is used, the model might mistakenly focus on irrelevant background details like bicycles or grassy areas. The only way to resolve this issue and improve performance is by incorporating additional annotation data, such as key points or human body parsing, to aid the ReID task.

Investigation on convergence and the average slope change. We studied the convergence of LFS-ReID and the evolution of the average slope during training. As seen in Fig. 5.6a, the distribution parameters $\boldsymbol{\mu}$ converge to specific values, showing our method's training stability. Figure 5.6b presents the average slope of the linear function in each interval, calculated from 500 randomly sampled loss functions using the current distribution parameters. This shows that LFS-ReID can dynamically adjust the gradient for samples of varying difficulty, which enhances the discriminative power of the learned features.

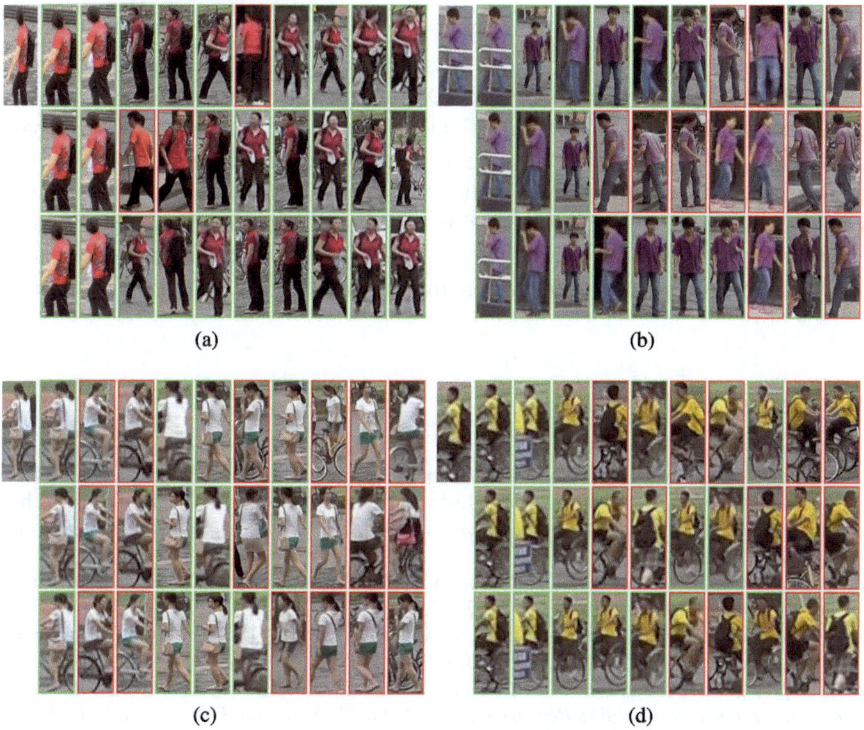

Fig. 5.5 An investigation of the retrieval results on the Market-1501 dataset using the MGN model. The top-left image is the query, and the next ten columns display the top ten retrieval results from the gallery. The three rows show the retrieval results for AM-LFS, SS, and our method. Green boxes indicate images with the same label as the query, while red boxes denote mismatched labels

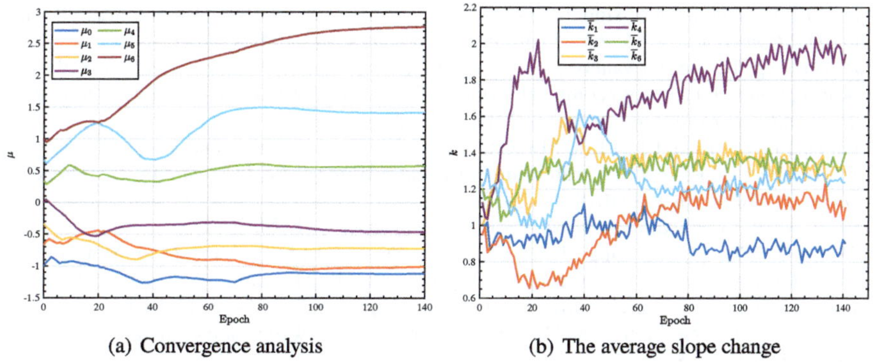

Fig. 5.6 An analysis of the convergence and average slope changes on the Market-1501 dataset, using the ResNet50 baseline model

5.5 Conclusion

In this chapter, we introduced LFS-ReID, an AutoML approach for loss function search specifically for person re-identification within the framework of margin-based softmax loss. We designed a sampling distribution based on non-independent truncated Gaussian distributions to ensure that the sampled loss function satisfies four key properties that we derived from analyzing margin-based softmax loss. Experimental results show our method achieves state-of-the-art performance on three datasets. We plan to extend our method to other domains, such as face recognition, in the future.

References

Adam Paszke, Sam Gross, Francisco Massa, Adam Lerer, James Bradbury, Gregory Chanan, Trevor Killeen, Zeming Lin, Natalia Gimelshein, Luca Antiga, et al. Pytorch: An imperative style, high-performance deep learning library. In *Advances in neural information processing systems*, pages 8026–8037, 2019.

Alexander Hermans, Lucas Beyer, and Bastian Leibe. In defense of the triplet loss for person re-identification. *CoRR*, abs/1703.07737, 2017.

Benoît Colson, Patrice Marcotte, and Gilles Savard. An overview of bilevel optimization. *Annals of operations research*, 153(1):235–256, 2007.

Barret Zoph and Quoc V. Le. Neural architecture search with reinforcement learning. In *5th International Conference on Learning Representations, ICLR 2017, Toulon, France, April 24–26, 2017, Conference Track Proceedings*. OpenReview.net, 2017.

Carl B Boyer and Uta C Merzbach. *A history of mathematics*. John Wiley & Sons, 2011.

Chao-Yuan Wu, R Manmatha, Alexander J Smola, and Philipp Krahenbuhl. Sampling matters in deep embedding learning. In *Proceedings of the IEEE International Conference on Computer Vision*, pages 2840–2848, 2017.

Chen Huang, Shuangfei Zhai, Walter Talbott, Miguel Bautista Martin, Shih-Yu Sun, Carlos Guestrin, and Josh Susskind. Addressing the loss-metric mismatch with adaptive loss alignment. In *International Conference on Machine Learning*, pages 2891–2900. PMLR, 2019.

Christian Szegedy, Vincent Vanhoucke, Sergey Ioffe, Jon Shlens, and Zbigniew Wojna. Rethinking the inception architecture for computer vision. In *Proceedings of the IEEE conference on computer vision and pattern recognition*, pages 2818–2826, 2016.

Chuming Li, Xin Yuan, Chen Lin, Minghao Guo, Wei Wu, Junjie Yan, and Wanli Ouyang. Am-lfs: Automl for loss function search. In *Proceedings of the IEEE International Conference on Computer Vision*, pages 8410–8419, 2019.

Curtis R Vogel. *Computational methods for inverse problems*. SIAM, 2002.

Di Wu, Si-Jia Zheng, Xiao-Ping Zhang, Chang-An Yuan, Fei Cheng, Yang Zhao, Yong-Jun Lin, Zhong-Qiu Zhao, Yong-Li Jiang, and De-Shuang Huang. Deep learning-based methods for person re-identification: A comprehensive review. *Neurocomputing*, 337:354–371, 2019.

Dirk P Kroese, Thomas Taimre, and Zdravko I Botev. *Handbook of monte carlo methods*, volume 706. John Wiley & Sons, 2013.

Ekin D Cubuk, Barret Zoph, Dandelion Mane, Vijay Vasudevan, and Quoc V Le. Autoaugment: Learning augmentation strategies from data. In *Proceedings of the IEEE conference on computer vision and pattern recognition*, pages 113–123, 2019.

Feng Wang, Jian Cheng, Weiyang Liu, and Haijun Liu. Additive margin softmax for face verification. *IEEE Signal Processing Letters*, 25(7):926–930, 2018.

Feng Zheng, Cheng Deng, Xing Sun, Xinyang Jiang, Xiaowei Guo, Zongqiao Yu, Feiyue Huang, and Rongrong Ji. Pyramidal person re-identification via multi-loss dynamic training. In *Proceedings of the IEEE Conference on Computer Vision and Pattern Recognition*, pages 8514–8522, 2019.

Guanshuo Wang, Yufeng Yuan, Xiong Chen, Jiwei Li, and Xi Zhou. Learning discriminative features with multiple granularities for person re-identification. In *Proceedings of the 26th ACM international conference on Multimedia*, pages 274–282, 2018.

Hanxiao Liu, Karen Simonyan, and Yiming Yang. DARTS: differentiable architecture search. In *7th International Conference on Learning Representations, ICLR 2019, New Orleans, LA, USA, May 6–9, 2019*. OpenReview.net, 2019.

Hao Luo, Wei Jiang, Youzhi Gu, Fuxu Liu, Xingyu Liao, Shenqi Lai, and Jianyang Gu. A strong baseline and batch normalization neck for deep person re-identification. *IEEE Trans. Multim.*, 22(10):2597–2609, 2020.

Hao Luo, Youzhi Gu, Xingyu Liao, Shenqi Lai, and Wei Jiang. Bag of tricks and a strong baseline for deep person re-identification. In *Proceedings of the IEEE Conference on Computer Vision and Pattern Recognition Workshops*, pages 0–0, 2019.

Hieu Pham, Melody Y. Guan, Barret Zoph, Quoc V. Le, and Jeff Dean. Efficient neural architecture search via parameter sharing. In Jennifer G. Dy and Andreas Krause, editors, *Proceedings of the 35th International Conference on Machine Learning, ICML 2018, Stockholmsmässan, Stockholm, Sweden, July 10–15, 2018*, volume 80 of *Proceedings of Machine Learning Research*, pages 4092–4101. PMLR, 2018.

Hao Wang, Yitong Wang, Zheng Zhou, Xing Ji, Dihong Gong, Jingchao Zhou, Zhifeng Li, and Wei Liu. Cosface: Large margin cosine loss for deep face recognition. In *Proceedings of the IEEE Conference on Computer Vision and Pattern Recognition*, pages 5265–5274, 2018.

Hengheng Zhang, Ying Li, Zijie Zhuang, Lingxi Xie, and Qi Tian. 3d-gat: 3d-guided adversarial transform network for person re-identification in unseen domains. *Pattern Recognition*, 112:107799, 2021.

Jiankang Deng, Jia Guo, Niannan Xue, and Stefanos Zafeiriou. Arcface: Additive angular margin loss for deep face recognition. In *Proceedings of the IEEE Conference on Computer Vision and Pattern Recognition*, pages 4690–4699, 2019.

Kaiming He, Xiangyu Zhang, Shaoqing Ren, and Jian Sun. Deep residual learning for image recognition. In *Proceedings of the IEEE conference on computer vision and pattern recognition*, pages 770–778, 2016.

Kaiyang Zhou, Yongxin Yang, Andrea Cavallaro, and Tao Xiang. Omni-scale feature learning for person re-identification. In *Proceedings of the IEEE International Conference on Computer Vision*, pages 3702–3712, 2019.

Karsten Roth, Timo Milbich, and Bjorn Ommer. Pads: Policy-adapted sampling for visual similarity learning. In *Proceedings of the IEEE/CVF Conference on Computer Vision and Pattern Recognition*, pages 6568–6577, 2020.

Keyu Tian, Chen Lin, Ming Sun, Luping Zhou, Junjie Yan, and Wanli Ouyang. Improving auto-augment via augmentation-wise weight sharing. *Advances in Neural Information Processing Systems*, 33, 2020.

Longhui Wei, Shiliang Zhang, Wen Gao, and Qi Tian. Person transfer gan to bridge domain gap for person re-identification. In *Proceedings of the IEEE conference on computer vision and pattern recognition*, pages 79–88, 2018.

Liang Zheng, Liyue Shen, Lu Tian, Shengjin Wang, Jingdong Wang, and Qi Tian. Scalable person re-identification: A benchmark. In *Proceedings of the IEEE international conference on computer vision*, pages 1116–1124, 2015.

Man Zhang, Yong Zhou, Jiaqi Zhao, Shixiong Xia, Jiaqi Wang, and Zizheng Huang. Semi-supervised blockwisely architecture search for efficient lightweight generative adversarial network. *Pattern Recognition*, 112:107794, 2021.

Nicolas Chopin. Fast simulation of truncated gaussian distributions. *Statistics and Computing*, 21(2):275–288, 2011.

Philipp Moritz, Robert Nishihara, Stephanie Wang, Alexey Tumanov, Richard Liaw, Eric Liang, Melih Elibol, Zongheng Yang, William Paul, Michael I. Jordan, and Ion Stoica. Ray: A distributed framework for emerging AI applications. In Andrea C. Arpaci-Dusseau and Geoff Voelker, editors, *13th USENIX Symposium on Operating Systems Design and Implementation, OSDI 2018, Carlsbad, CA, USA, October 8–10, 2018*, pages 561–577. USENIX Association, 2018.

Ronald J Williams. Simple statistical gradient-following algorithms for connectionist reinforcement learning. *Machine learning*, 8(3-4):229–256, 1992.

Ruibing Hou, Bingpeng Ma, Hong Chang, Xinqian Gu, Shiguang Shan, and Xilin Chen. Interaction-and-aggregation network for person re-identification. In *Proceedings of the IEEE Conference on Computer Vision and Pattern Recognition*, pages 9317–9326, 2019.

Ruijie Quan, Xuanyi Dong, Yu Wu, Linchao Zhu, and Yi Yang. Auto-reid: Searching for a part-aware convnet for person re-identification. In *Proceedings of the IEEE International Conference on Computer Vision*, pages 3750–3759, 2019.

Sergey Ioffe and Christian Szegedy. Batch normalization: Accelerating deep network training by reducing internal covariate shift. In Francis R. Bach and David M. Blei, editors, *Proceedings of the 32nd International Conference on Machine Learning, ICML 2015, Lille, France, 6–11 July 2015*, volume 37 of *JMLR Workshop and Conference Proceedings*, pages 448–456. JMLR.org, 2015.

Sungbin Lim, Ildoo Kim, Taesup Kim, Chiheon Kim, and Sungwoong Kim. Fast autoaugment. In *Advances in Neural Information Processing Systems*, pages 6665–6675, 2019.

Tianlong Chen, Shaojin Ding, Jingyi Xie, Ye Yuan, Wuyang Chen, Yang Yang, Zhou Ren, and Zhangyang Wang. Abd-net: Attentive but diverse person re-identification. In *Proceedings of the IEEE International Conference on Computer Vision*, pages 8351–8361, 2019.

Wanyin Wu, Dapeng Tao, Hao Li, Zhao Yang, and Jun Cheng. Deep features for person re-identification on metric learning. *Pattern Recognition*, 110:107424, 2021.

Wei Li, Rui Zhao, Tong Xiao, and Xiaogang Wang. Deepreid: Deep filter pairing neural network for person re-identification. In *Proceedings of the IEEE conference on computer vision and pattern recognition*, pages 152–159, 2014.

Wei Li, Xiatian Zhu, and Shaogang Gong. Harmonious attention network for person re-identification. In *Proceedings of the IEEE conference on computer vision and pattern recognition*, pages 2285–2294, 2018.

Weiyang Liu, Yandong Wen, Zhiding Yu, Ming Li, Bhiksha Raj, and Le Song. Sphereface: Deep hypersphere embedding for face recognition. In *Proceedings of the IEEE conference on computer vision and pattern recognition*, pages 212–220, 2017.

Xing Fan, Wei Jiang, Hao Luo, and Mengjuan Fei. Spherereid: Deep hypersphere manifold embedding for person re-identification. *Journal of Visual Communication and Image Representation*, 60:51–58, 2019.

Xiaobo Wang, Shuo Wang, Cheng Chi, Shifeng Zhang, and Tao Mei. Loss function search for face recognition. In *Proceedings of the 37th International Conference on Machine Learning, ICML 2020, 13–18 July 2020, Virtual Event*, volume 119 of *Proceedings of Machine Learning Research*, pages 10029–10038. PMLR, 2020.

Xinyu Zhang, Qiang Wang, Jian Zhang, and Zhao Zhong. Adversarial autoaugment. In *8th International Conference on Learning Representations, ICLR 2020, Addis Ababa, Ethiopia, April 26–30, 2020*. OpenReview.net, 2020.

Yandong Wen, Kaipeng Zhang, Zhifeng Li, and Yu Qiao. A discriminative feature learning approach for deep face recognition. In *European conference on computer vision*, pages 499–515. Springer, 2016.

Yang Fu, Yunchao Wei, Yuqian Zhou, Honghui Shi, Gao Huang, Xinchao Wang, Zhiqiang Yao, and Thomas Huang. Horizontal pyramid matching for person re-identification. In *Proceedings of the AAAI Conference on Artificial Intelligence*, volume 33, pages 8295–8302, 2019.

Yifan Sun, Changmao Cheng, Yuhan Zhang, Chi Zhang, Liang Zheng, Zhongdao Wang, and Yichen Wei. Circle loss: A unified perspective of pair similarity optimization. In *Proceedings of the IEEE/CVF Conference on Computer Vision and Pattern Recognition*, pages 6398–6407, 2020.

Yifan Sun, Liang Zheng, Weijian Deng, and Shengjin Wang. Svdnet for pedestrian retrieval. In *Proceedings of the IEEE International Conference on Computer Vision*, pages 3800–3808, 2017.

Yifan Sun, Liang Zheng, Yi Yang, Qi Tian, and Shengjin Wang. Beyond part models: Person retrieval with refined part pooling (and a strong convolutional baseline). In *Proceedings of the European Conference on Computer Vision (ECCV)*, pages 480–496, 2018.

Yong Wu, Kun Zhang, Di Wu, Chao Wang, Chang-An Yuan, Xiao Qin, Tao Zhu, Yu-Chuan Du, Han-Li Wang, and De-Shuang Huang. Person re-identification by multi-scale feature representation learning with random batch feature mask. *IEEE Transactions on Cognitive and Developmental Systems*, 2020.

Zhedong Zheng, Xiaodong Yang, Zhiding Yu, Liang Zheng, Yi Yang, and Jan Kautz. Joint discriminative and generative learning for person re-identification. In *Proceedings of the IEEE conference on computer vision and pattern recognition*, pages 2138–2147, 2019.

Zhun Zhong, Liang Zheng, Donglin Cao, and Shaozi Li. Re-ranking person re-identification with k-reciprocal encoding. In *Proceedings of the IEEE Conference on Computer Vision and Pattern Recognition*, pages 1318–1327, 2017.

Zhun Zhong, Liang Zheng, Guoliang Kang, Shaozi Li, and Yi Yang. Random erasing data augmentation. In *The Thirty-Fourth AAAI Conference on Artificial Intelligence, AAAI 2020, The Thirty-Second Innovative Applications of Artificial Intelligence Conference, IAAI 2020, The Tenth AAAI Symposium on Educational Advances in Artificial Intelligence, EAAI 2020, New York, NY, USA, February 7–12, 2020*, pages 13001–13008. AAAI Press, 2020.

Chapter 6
Static Generalized Margin-Based Softmax Loss Function Search for Person Re-identification

6.1 Introduction

Person re-identification (ReID), a pivotal task in computer vision, involves matching a specific individual across various surveillance cameras that do not share overlapping fields of view (Ye et al. 2021; Chen et al. 2021; Fan et al. 2019). Propelled by the advancements in deep learning and the growing demand for smart video surveillance, ReID has garnered significant attention within the research community. Despite the considerable progress made by deep neural network (DNN)-based models, the task of learning robust, discriminative features for accurate person identification in large-scale datasets remains a formidable challenge. This difficulty stems from substantial intra-class variations caused by factors such as changes in pose, occlusions, and cluttered backgrounds.

Recent advancements in person re-identification (ReID) have predominantly focused on innovating network architectures to extract discriminative features, often overlooking the crucial role of the loss function (Wang et al. 2018a; Zheng et al. 2019; Chen et al. 2021; He et al. 2021; Hongyang et al. 2021). It's widely recognized that achieving high performance with a deep neural network (DNN) model requires both an effective network design and a suitable loss function. However, most contemporary works still rely on the conventional combination of cross-entropy and triplet loss. The exploration of alternative loss functions for the ReID problem remains limited. Drawing inspiration from the NormFace approach (Wang et al. 2017) in face recognition (FR), Fan et al. (2019) introduced SphereReID to learn a hypersphere manifold embedding, which demonstrated an advantage over the standard cross-entropy loss in feature extraction. Yet, recent FR research indicates that NormFace is not the most effective among margin-based softmax (MS) losses (Liu et al. 2017; Wang et al. 2018b; Deng et al. 2019). Sun et al. (2020) later proposed CircleLoss (Sun et al. 2020), a more adaptable loss function that has shown superior performance in both FR and ReID tasks compared to other MS losses. The aforementioned loss functions are merely specific instances within the broader family of generalized margin-based

H. Gu et al., *Automated Machine Learning for Person Re-Identification*,
Intelligent Perception and Information Processing,
https://doi.org/10.1007/978-981-95-3433-3_6

softmax (GMS) loss functions and likely represent only a fraction of the possibilities. It is probable that they are not the optimal choices within this expansive space. Consequently, this chapter seeks to investigate whether a more effective loss function exists within the GMS family, surpassing the performance of these established methods.

The rise of automated machine learning (AutoML) has led to the proposal of numerous automated methods, particularly in domains such as data augmentation (Cubuk et al. 2019; Lim et al. 2019; Zhang et al. 2020; Tian et al. 2020) and network architecture design (Zoph and Le 2017; Liu et al. 2019; Zhang et al. 2021), where they have surpassed traditional hand-crafted techniques. In the realm of loss functions, AutoML-based approaches have also recently emerged. Current methods for searching loss functions primarily fall into two categories: (1) The first method, similar to Population-Based Training (PBT) (Jaderberg et al. 2017), does not directly optimize the loss function's specific form but rather focuses on the performance of the final trained network (Li et al. 2019; Wang et al. 2020a). This approach necessitates re-running the computationally expensive search process every time the dataset or network architecture is modified. (2) The second method, which is the focus of our work, involves searching for a loss function composed of predefined primitive operations (Liu et al. 2021; Li et al. 2021a) or a parameterized metric-surrogate loss (Li et al. 2021b). A significant advantage of this second approach is that the discovered loss function can be directly transferred to similar tasks without requiring another extensive search. However, a major challenge with this method, as with many other AutoML techniques, is the high computational cost associated with the search. While several techniques have been proposed to accelerate this process (Liu et al. 2021; Li et al. 2021a), the gains in search efficiency remain modest, indicating a need for further improvement.

This chapter introduces AutoLoss-GMS, a novel methodology for automatically discovering optimal loss functions within the generalized margin-based softmax (GMS) loss space for person re-identification (ReID). Our approach builds upon the second search method previously discussed. Specifically, we represent the GMS loss function using a novel paradigm involving two computational graphs and a constant value. To efficiently navigate this vast search space, we employ a sophisticated evolutionary algorithm. The computational graphs are built using a forward-construction method, which generates a more diverse range of loss function forms compared to the backward method used in prior works (Liu et al. 2021; Li et al. 2021a). To enhance the diversity of the generated loss functions, we introduce a cross-graph mutation operation in addition to standard in-graph mutations. To address the computational cost inherent in AutoML methods, we incorporate several strategies to boost search efficiency. These include a loss-rejection protocol, an equivalence-check strategy, and a predictor-based promising-loss chooser, which is, to our knowledge, the first time a predictor has been used in loss function search (LFS).

The key contributions of this work are summarized as follows:

- We present AutoLoss-GMS, the pioneering work to automatically search for GMS loss functions specifically tailored for person ReID.

- We propose an efficient general search framework that incorporates a performance predictor to accelerate the discovery of effective GMS loss functions.
- Our discovered loss functions are highly transferable, achieving competitive, and in some cases, state-of-the-art performance on various ReID models and datasets.

6.2 Related Works

6.2.1 Person Re-identification

Person re-identification (ReID) has recently gained significant attention in the computer vision community. With the advent of deep neural networks (DNNs), the performance of DNN-based models has surpassed human capabilities. The core challenge in ReID is to extract discriminative features. While most existing research has focused on designing novel network architectures to enhance feature discrimination (Wang et al. 2018a; Chen et al. 2019; Sun et al. 2018), the loss function also plays a critical role in this process. In the ReID domain, cross-entropy and triplet loss are the most common paradigms, with very few works exploring alternative loss function forms. Inspired by the NormFace method (Wang et al. 2017) from face recognition (FR), Fan et al. (2019) introduced SphereReID, which experimentally demonstrated that features embedded on a hypersphere are more discriminative than those learned with cross-entropy. However, SphereReID only considers hyperspherical embedding and neglects the benefits of a large margin, which has been shown in FR to outperform NormFace (Liu et al. 2017; Wang et al. 2018b; Deng et al. 2019). Sun et al. (2020) addressed this by proposing a more flexible loss called CircleLoss, which showed significant improvement over other margin-based softmax (MS) loss functions. Designing such effective loss functions manually is a difficult and time-consuming process that requires extensive domain expertise. To alleviate this burden, methods like AM-LFS (Li et al. 2019) have been proposed to automatically design loss functions within the MS space. Nevertheless, AM-LFS does not produce a loss function with a fixed, explicit form, which makes it challenging to directly transfer the discovered loss to different datasets or networks without an additional search.

6.2.2 Loss Function Search

The loss function is a crucial component of deep learning, and its manual design often demands significant domain expertise. Consequently, Loss Function Search (LFS) via automated machine learning (AutoML) has become a sought-after objective. Existing methods like AM-LFS (Li et al. 2019) and Searched-Softmax (Wang et al. 2020b) employ reinforcement learning to dynamically find the function $t(x)$ within the margin-based softmax (MS) loss framework. A major drawback of these approaches is that their outputs are dynamic policies rather than fixed, transferable expressions, which necessitates re-running the expensive search process for every new dataset or network architecture. In contrast, other LFS methods, more akin to Network Architecture Search (NAS), are capable of producing a fixed-form result. Auto Seg-Loss (Li et al. 2021b) searches for a differentiable approximation of non-differentiable metrics, but this method is not suitable for our specific search space. AutoLoss-Zero (Li et al. 2021a) and CSE-Autoloss (Liu et al. 2021) represent the loss function as a computational graph built from basic mathematical operations. This representation is ideal for our goal of finding a fixed-form loss function. However, these methods suffer from a high computational cost. Although they introduce various techniques to improve search efficiency, the gains remain limited, highlighting a persistent challenge in this area.

6.3 Methodology

In this section, we present a detailed overview of our proposed method, encompassing the design of the search space and the search algorithm.

6.3.1 *Search Space*

6.3.1.1 Preliminary Knowledge

To clearly define our search space, we first introduce some preliminary concepts. We begin by considering the standard softmax cross-entropy loss:

$$\mathcal{L}_{ce} = -\log\left(\frac{\exp\left(\boldsymbol{W}_y^{\mathsf{T}}\boldsymbol{x} + b_y\right)}{\sum_{i=1}^{K}\exp\left(\boldsymbol{W}_i^{\mathsf{T}}\boldsymbol{x} + b_i\right)}\right), \tag{6.1}$$

where $\boldsymbol{x} \in \mathbb{R}^d$ is the feature vector, $y \in \{1, \ldots, K\}$ its ground-truth label, K the class count, and $\boldsymbol{W}_i, b_i$ the parameters of class i. Omitting biases, enforcing unit norms on both classifier weights and features (i.e., $\|\boldsymbol{W}_i\| = \|\boldsymbol{x}\| = 1$), and introducing a positive scale s yields the unified margin-based softmax (MS) loss:

$$\mathcal{L}_{ms} = -\log\left(\frac{\exp\left(s \cdot t\left(\cos\theta_y\right)\right)}{\exp\left(s \cdot t\left(\cos\theta_y\right)\right) + \sum_{i\neq y}\exp\left(s \cdot \cos\theta_i\right)}\right), \tag{6.2}$$

with $\cos\theta_i = \boldsymbol{W}_i^{\mathsf{T}}\boldsymbol{x}$ and $t : [-1, 1] \to \mathbb{R}$. Generalizing further by replacing the non-target logits with an arbitrary mapping $n(\cdot)$ gives the GMS loss:

Table 6.1 The $t(x)$ and $n(x)$ of the hand-crafted loss functions. The de(x) represents the "Detach" operation

Loss	$t(x)$	$n(x)$
NormFace	x	x
CosFace	$x - m$	x
ArcFace	$\cos(\arccos(x) + m)$	x
SphereFace	$\cos(m \arccos(x))$	x
CircleLoss	$[\mathrm{de}(1 + m - x)]_+(x - 1 + m)$	$[\mathrm{de}(m + x)]_+(x - m)$

$$\mathcal{L}_{gms} = -\log\left(\frac{\exp\left(s \cdot t\left(\cos\theta_y\right)\right)}{\exp\left(s \cdot t\left(\cos\theta_y\right)\right) + \sum_{i \neq y} \exp\left(s \cdot n\left(\cos\theta_i\right)\right)}\right). \tag{6.3}$$

Evidently, $\mathcal{L}_{ms}$ is recovered when $n(x) = x$. Table 6.1 shows that prevalent MS variants (Wang et al. 2017; Liu et al. 2017; Wang et al. 2018b; Deng et al. 2019) and CircleLoss (Sun et al. 2020) can be instantiated by selecting appropriate $t(x)$ and $n(x)$. The goal of this chapter is to ascertain the optimal triple $\Theta = \{s, t(x), n(x)\}$ under $\mathcal{L}_{gms}$, cast as the following bi-level program:

$$\begin{aligned} \Theta^* &= \arg\max_{\Theta} \;\; \xi\left(\mathcal{M}_{\Omega^*(\Theta)}; \mathcal{D}_{val}\right), \\ \text{s.t.}\Omega^*\left(\Theta\right) &= \arg\min_{\Omega} \;\; \mathbb{E}_{(\boldsymbol{x},y)\in\mathcal{D}_{trn}}\left[\mathcal{L}^{\Theta}_{gms}\left(\mathcal{M}_{\Omega}\left(\boldsymbol{x}\right), y\right)\right], \end{aligned} \tag{6.4}$$

where $\mathcal{M}_{\Omega}$ denotes a network with parameters Ω, $\mathcal{D}_{trn}$ and $\mathcal{D}_{val}$ are the training and validation sets, ξ is the evaluation metric, and $\mathcal{L}^{\Theta}_{gms}$ signifies the GMS loss parameterized by Θ.

6.3.1.2 Loss Function Representation

The goal is to find the optimal parameter set $\Theta = \{s, t(x), n(x)\}$. Since these three components are independent in the GMS loss function $\mathcal{L}_{gms}$, we can model them separately, as shown in Fig. 6.1 and Table 6.2.

The scale factor s is defined as $s \geq 1$, which is consistent with existing manually designed loss functions. We discretize s into a finite set of values:

$$s \in Scales = \left\{2^{\Delta_s \times i} | i \in \mathbb{N}, 0 \leq i \leq N_s\right\}, \tag{6.5}$$

where $\Delta_s > 0$ and N_s are predetermined constants. This simple discretization has been shown to be effective in our experiments.

The functions $t(x)$ and $n(x)$ are represented as computational graphs (CGs). We'll detail the representation for $t(x)$, as $n(x)$ is represented similarly. A computational graph $\mathcal{G}_t$ for $t(x)$ has three types of nodes:

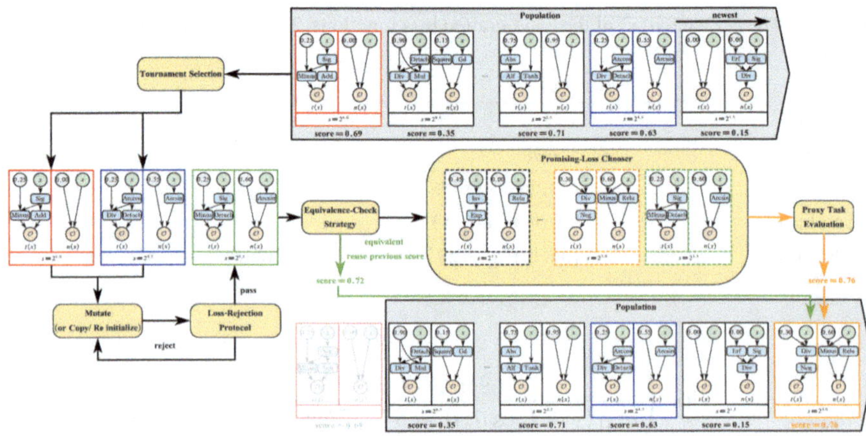

Fig. 6.1 The search pipeline of AutoLoss-GMS

1. Input Nodes: There are two types of input nodes:
   - Variable Inputs: A single node, represented as x (the green node in Fig. 6.2), which corresponds to the network's output $\cos\theta_y$ from the $\mathcal{L}_{gms}$ equation.
   - Constant Inputs: A set of predefined constant values, c (the gray nodes in Fig. 6.2). These constants are also discretized as:

$$c \in \text{Cons} = \{\Delta_c \times i \mid i \in \mathbb{N}, 0 \leq i \leq N_c\},$$

   where $\Delta_c > 0$ and N_c are predefined constants.

2. Intermediate Nodes: These nodes (the blue nodes in Fig. 6.2) perform primitive mathematical operations selected from a predefined set $\mathcal{H}$ (e.g., addition and multiplication).
3. Output Node: The final node (the orange node O in Fig. 6.2), which produces the final value of $t(x)$.

In summary, our search space is defined by the triple $\Theta = \{s, \mathcal{G}_t, \mathcal{G}_n\}$, where s is chosen from the discrete set Scales, and $\mathcal{G}_t$ and $\mathcal{G}_n$ are the computational graphs for $t(x)$ and $n(x)$, respectively.

6.3.2 *Search Algorithm*

Our search algorithm is primarily a variant of evolutionary algorithms (Real et al. 2019, 2020; Li et al. 2021a), which is well-suited for parallelization. The search pipeline, shown in Fig. 4, begins with the Initialization of an initial population of K loss functions. In each evolutionary step, two parent loss functions are chosen from

Table 6.2 The set of primitive operations $\mathcal{H}$ used in our computational graph search

Operator	Expression	Arity
Add	$x + y$	2
Minus	$x - y$	2
Mul	$x \times y$	2
Div	x/y	2
Neg	$-x$	1
Abs	$\|x\|$	1
Inv	$1/x$	1
Log	$\log(x)$	1
Exp	e^x	1
Sig	$1/(1+e^{-x})$	1
Tanh	$(e^x - e^{-x})/(e^x + e^{-x})$	1
Sin	$\sin(x)$	1
Cos	$\cos(x)$	1
Arcsin	$\arcsin(x)$	1
Arccos	$\arccos(x)$	1
Square	x^2	1
Sqrt	$\sqrt{x}$	1
Relu	$[x]_+$	1
Softplus	$\log(1+e^x)$	1
Gd	$2\arctan((e^{x/2} - e^{-x/2})/(e^{x/2} + e^{-x/2}))$	1
Alf	$x/\sqrt{1+x^2}$	1
Erf	$\frac{2}{\sqrt{\pi}} \int_0^x e^{-t^2} dt$	1
Erfc	$1 - \frac{2}{\sqrt{\pi}} \int_0^x e^{-t^2} dt$	1
Detach	$\mathrm{de}(x)$	1

the current population using two independent tournament selections (Mirjalili 2019). Each tournament selects a parent from a random subset of the population (with a size determined by the T ratio). These selected parents then undergo a series of well-designed mutation operations to generate new offspring. The offspring are created and mutated until they pass our loss-rejection protocol, which filters out invalid or ill-formed functions.

To improve efficiency, we use an equivalence-check strategy based on a feature vector. This strategy accurately identifies and discards mathematically equivalent loss functions, preventing redundant evaluations. From the remaining non-equivalent candidates, the predictor-based promising-loss chooser selects the most promising-loss functions for full evaluation. Following previous work (Li et al. 2021a; Real et al. 2020), we maintain only the most recent P loss functions in the population to manage its size.

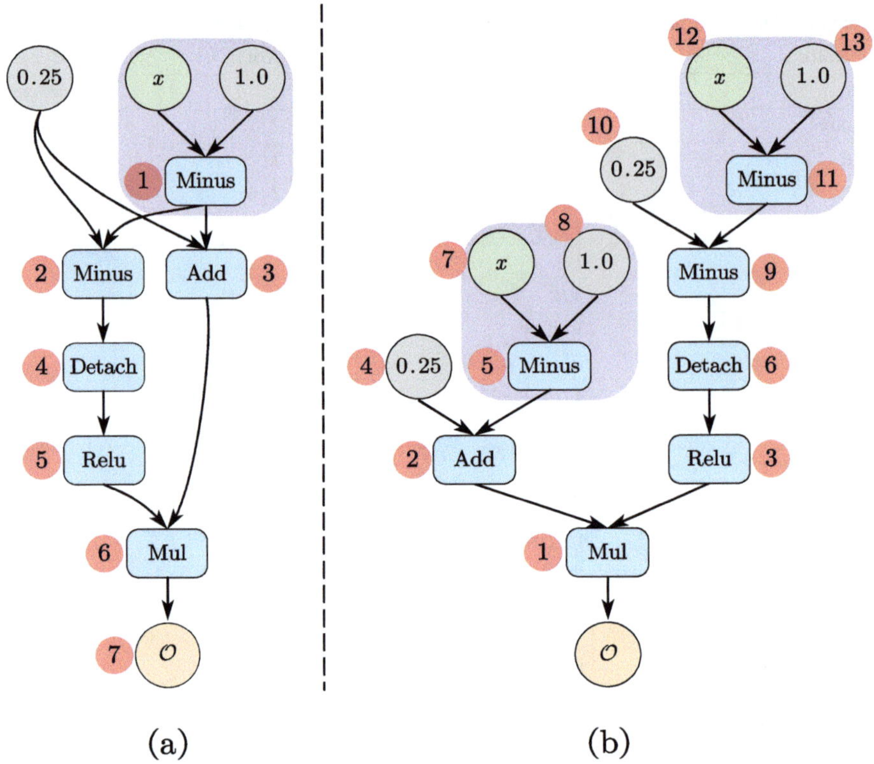

Fig. 6.2 Computational Graph $\mathcal{G}_t$ for the $t(x)$ component of CircleLoss. This figure illustrates two different construction methods for the same function, where $m = 0.25$. Graph (a) is built using a forward method, showing the order of operations as they are executed from inputs to output. In contrast, graph (b) is built using a backward method, which starts from the output and works its way back to the inputs. The numbers within the red nodes indicate the sequence in which each node is added during construction

6.3.2.1 Initialization and Mutation

To initialize the population, we adopt two distinct approaches: (1) **Randomly generated loss functions** for de novo search, similar to AutoLoss-Zero (Li et al. 2021a); (2) **Predefined hand-crafted loss functions** leveraging prior knowledge, akin to CSE-Autoloss (Liu et al. 2021). The predefined hand-crafted loss functions are primarily derived from Table 6.1.

AutoLoss-Zero and CSE-Autoloss employ a backward construction method for computational graphs (CGs) (Li et al. 2021a; Liu et al. 2021), which is less efficient. Specifically, intermediate node outputs cannot be reused multiple times, as illustrated in the purple section of Fig. 6.2b. To enhance efficiency, we propose a forward-construction method inspired by neural architecture search (NAS) (Zoph and Le 2017).

The forward-construction of a CG begins with $1 + n_c$ input nodes: one feature node x (green in Fig. 6.2) and n_c constant nodes (gray in Fig. 6.2). Given the current state of the CG, all feasible operations and edge connections are enumerated, and a random operation along with its corresponding edge is selected and added to the CG. When the chosen operation is "Output", all nodes without successors are connected to the output node $\mathcal{O}$. Once the CG reaches the maximum number of primitive operations n_o, only the "Output" operation can be selected. Compared to the backward method, the forward method can generate a much richer variety of loss function forms with the same number of primitive operations. The forward-construction process of a CG is detailed in Algorithm 6.1, and the construction of our search space's loss function is outlined in Algorithm 6.2.

Algorithm 6.1 Construct_CG

Require: The value set of constant nodes $Cons$, the number of constant nodes n_c, the maximum number of primitive operations n_o, and the primitive operations set $\mathcal{H}$.
Ensure: The computational graph $\mathcal{G}(\mathcal{V}, \mathcal{E})$.

$\mathcal{V} \leftarrow [x]$ ▷ Initialize nodes set with input variable
$\mathcal{E} \leftarrow [\,]$ ▷ Initialize edges set as empty
for $i \leftarrow 1$ to n_c **do**
    Sample con_i from $Cons$
    Append con_i to $\mathcal{V}$
end for
while $|\mathcal{V}| \leq n_o + n_c + 1$ **do**
    Enumerate the possible set $\{(Op_i, E_i)\}$ based on $\mathcal{G}(\mathcal{V}, \mathcal{E})$
    if $|\mathcal{V}| = n_o + n_c + 1$ **then**
        $Op \leftarrow \mathcal{O}$ ▷ Set the final operation as the output node
        $E \leftarrow \{e_{v \to \mathcal{O}} \mid v \in \mathcal{V}$ and v has no successor$\}$
    else
        Sample (Op, E) from $\{(Op_i, E_i)\}$
    end if
    Append Op to $\mathcal{V}$
    for each e in E **do**
        Append e to $\mathcal{E}$
    end for
end while
return $\mathcal{G}(\mathcal{V}, \mathcal{E})$

Algorithm 6.2 Construct_LF

Require: The value set of the scale factor $Scales$, the value set of constant nodes $Cons$, the number of constant nodes n_c, the maximum number of primitive operations n_o, and the primitive operations set $\mathcal{H}$.
Ensure: The loss function $\mathcal{L}_g$ parameterized by $\Theta = \{s, \mathcal{G}_t(\mathcal{V}_t, \mathcal{E}_t), \mathcal{G}_n(\mathcal{V}_n, \mathcal{E}_n)\}$.

Sample s from $Scales$
$\mathcal{G}_t(\mathcal{V}_t, \mathcal{E}_t) \leftarrow$ Construct_CG($Cons, n_c, n_o, \mathcal{H}$)
$\mathcal{G}_n(\mathcal{V}_n, \mathcal{E}_n) \leftarrow$ Construct_CG($Cons, n_c, n_o, \mathcal{H}$)
return $\{s, \mathcal{G}_t(\mathcal{V}_t, \mathcal{E}_t), \mathcal{G}_n(\mathcal{V}_n, \mathcal{E}_n)\}$

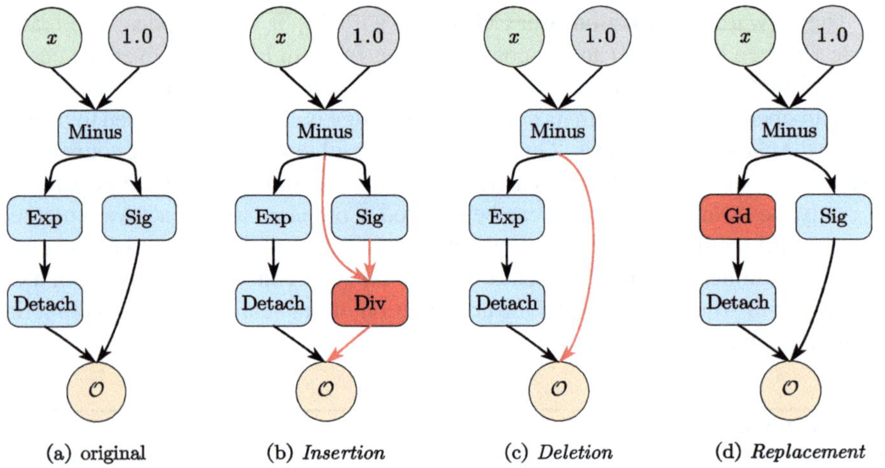

Fig. 6.3 Illustration of the three in-graph mutation operations. The original Computational Graph (CG) is shown in (a). The red parts in (b), (c), and (d) highlight the changes made by each mutation. Graph (b) shows an "Insertion" of a new operation. Graph (c) demonstrates a "Deletion" of an existing node. Finally, Graph (d) illustrates a "Replacement" of a node with a different operation

In addition to the three in-graph mutation operations (Li et al. 2021a), which are shwon as follows:

- *Insertion*. An operation randomly sampled from $\mathcal{H}$ is inserted between a randomly selected non-root node and its parent. When the selected operation's arity is 2, it would randomly select one node from all nodes generated before the selected non-root node as the other parent node. If the CG has reached the maximum number of primitive operations n_o, this mutation is valid.
- *Deletion*. Any intermediate computational node can be selected and removed. When the removed operation's arity is 2, the randomly selected parent node becomes the new parent node of the removed node' all child nodes.
- *Replacement*. The randomly selected non-root node can be replaced by an operation randomly selected from $\mathcal{H}$. If the selected non-root node' arity is smaller than that of the selected operation, it would randomly select one node from all nodes generated before the selected non-root node as the other parent node. Otherwise, the randomly selected parent node from the parent nodes of the selected non-root node would be selected as the parent of the selected operation.

Figure 6.3 shows examples of these in-graph mutation operations.

We also introduce a cross-graph mutation operation. Specifically, in each evolution iteration, given two loss functions ($\Theta^{(1)}$ and $\Theta^{(2)}$), $\mathcal{G}_t$ or $\mathcal{G}_n$ in $\Theta^{(2)}$ replaces the corresponding computational graph in $\Theta^{(1)}$ with a probability of p_c to obtain the mutated Θ'. The scale factor s' in Θ' is updated as follows:

$$s' = 2^{[(\log_2(s^{(1)})+\log_2(s^{(2)}))/2|\Delta_s]}, \tag{6.6}$$

where $[x|\Delta_s]$ denotes standardizing the precision of x to Δ_s. Given the significant impact of hyperparameters in the loss function on the results, we also introduce mutation operations on s and the constant terms $con_i|_{i=1}^{n_c}$, rather than keeping these constants fixed as in AutoLoss-Zero and CSE-Autoloss. The specific mutation operations for the constant terms are defined as:

$$\begin{cases} s' = 2^{Clamp\left[\log_2(s)+\Delta_s r;0;\Delta_s N_s\right]} \\ c' = Clamp\left[c + \Delta_c r;\, 0;\, \Delta_c N_c\right] \end{cases}, \tag{6.7}$$

where $Clamp[x; x_{min}; x_{max}]$ clamps x within the range $[x_{min}, x_{max}]$, and r is randomly selected from $\{-1, 0, 1\}$.

Given the two loss functions $\Theta^{(1)}$ and $\Theta^{(2)}$ selected via tournament, the offspring is produced according to Algorithm 6.3.

6.3.2.2 Loss-Rejection Protocol

Drawing inspiration from CES-Autoloss (Liu et al. 2021) and AutoLoss-Zero (Li et al. 2021a), our loss-rejection protocol is initiated from two distinct perspectives:

Basic properties protocol. Upon analyzing the GMS loss function $\mathcal{L}_{gms}$, the following properties are typically required to be satisfied:

- $t'(x) \geq 0, x \in [-1, 1]$;
- $n'(x) \geq 0, x \in [-1, 1]$;
- $n(x) - t(x) \geq 0, x \in [-1, 1]$.

Firstly, the gradients of $\mathcal{L}_{gms}$ with respect to $\cos\theta_y$ and $\cos\theta_i (i \neq y)$ can be obtained:

$$\begin{cases} \frac{\partial \mathcal{L}_{gms}}{\partial \cos\theta_y} = -s \cdot t'\left(\cos\theta_y\right) \cdot \left(1 - p_y^{t,n}\right) \\ \frac{\partial \mathcal{L}_{gms}}{\partial \cos\theta_i} = s \cdot n'\left(\cos\theta_i\right) \cdot \left(1 - p_i^{t,n}\right), i \neq y \end{cases}, \tag{6.8}$$

where

$$\begin{cases} p_y^{t,n} = \frac{\exp\left(s \cdot t\left(\cos\theta_y\right)\right)}{\exp\left(s \cdot t\left(\cos\theta_y\right)\right) + \sum_{i\neq y}\exp\left(s \cdot n(\cos\theta_i)\right)} \\ p_i^{t,n} = \frac{\exp\left(s \cdot n(\cos\theta_i)\right)}{\exp\left(s \cdot t\left(\cos\theta_y\right)\right) + \sum_{i\neq y}\exp\left(s \cdot n(\cos\theta_i)\right)}, i \neq y \end{cases}. \tag{6.9}$$

As the training progresses, the overall trend of $\cos\theta_y$ is increasing, while the overall trend of $\cos\theta_i (i \neq y)$ is decreasing. Therefore, from the overall trend, the following properties can be obtained:

$$\frac{\partial \mathcal{L}_{gms}}{\partial \cos\theta_y} \leq 0, \frac{\partial \mathcal{L}_{gms}}{\partial \cos\theta_i} \geq 0, i \neq y,$$

and $p_y^{t,n} < 1$ and $p_i^{t,n} < 1$, so we can obtain the following two properties in $\mathcal{L}_{gms}$:

Algorithm 6.3 Produce_Offspring

Require: Two parent loss functions $\Theta^{(1)}$ and $\Theta^{(2)}$, the crossover probability p_c, the copy probability p_{copy}, the re-initialization probability p_{re}, the maximum number of mutations N_m, the parameters Δ_c and N_c for $Cons$, the parameters Δ_s and N_s for $Scales$, the number of constant nodes n_c, the maximum number of primitive operations n_o and the primitive operations set $\mathcal{H}$.
Ensure: The offspring Θ'.

$\Theta^{(rand)} \leftarrow$ Construct_LF($Scales, Cons, n_c, n_o, \mathcal{H}$)
$\Theta' \leftarrow \Theta^{(1)}$
if $rand < p_c$ **then** ▷ Crossover $\Theta^{(1)}$ and $\Theta^{(2)}$
  if $rand < 0.5$ **then**
    $\Theta' \leftarrow \{s', \mathcal{G}_t^{(2)}, \mathcal{G}_n'\}$
  else
    $\Theta' \leftarrow \{s', \mathcal{G}_t', \mathcal{G}_n^{(2)}\}$
  end if
  Update the s' in Θ' using Eq. (6.6)
end if
$r \leftarrow rand$ ▷ Mutations on $\mathcal{G}_t'$
if $r < p_{copy}$ **then** ▷ Directly copy $\mathcal{G}_t'$
  Pass ▷ No operation
else if $r < p_{re}$ **then** ▷ Re-initialize $\mathcal{G}_t'$ with $\mathcal{G}_t^{rand}$
  $\Theta' \leftarrow \{s', \mathcal{G}_t^{rand}, \mathcal{G}_n'\}$
else
  $N \leftarrow$ randint($1, N_m$) ▷ Mutate $\mathcal{G}_t'$
  for $n \leftarrow 1$ to N **do**
    Mutate $con_i|_{i=1}^{n_c}$ in $\mathcal{G}_t'$ using Eq. (6.7)
    Mutate $\mathcal{G}_t'$ with a random operation from $\{Insertion, Deletion, Replacement\}$
  end for
end if
$r \leftarrow rand$ ▷ Mutations on $\mathcal{G}_n'$
if $r < p_{copy}$ **then** ▷ Directly copy $\mathcal{G}_n'$
  Pass ▷ No operation
else if $r < p_{re}$ **then** ▷ Re-initialize $\mathcal{G}_n'$ with $\mathcal{G}_n^{rand}$
  $\Theta' \leftarrow \{s', \mathcal{G}_t', \mathcal{G}_n^{rand}\}$
else
  $N \leftarrow$ randint($1, N_m$) ▷ Mutate $\mathcal{G}_n'$
  for $n \leftarrow 1$ to N **do**
    Mutate $con_i|_{i=1}^{n_c}$ in $\mathcal{G}_n'$ using Eq. (6.7)
    Mutate $\mathcal{G}_n'$ with a random operation from $\{Insertion, Deletion, Replacement\}$
  end for
end if
$r \leftarrow rand$ ▷ Mutations on s'
if $r < p_{copy}$ **then** ▷ Directly copy s'
  Pass ▷ No operation
else if $r < p_{re}$ **then** ▷ Re-initialize s' with s^{rand}
  $\Theta' \leftarrow \{s^{rand}, \mathcal{G}_t', \mathcal{G}_n'\}$
else
  $N \leftarrow$ randint($1, N_m$) ▷ Mutate on s'
  for $n \leftarrow 1$ to N **do**
    Mutate the $con_i|_{i=1}^{n_c}$ in $\mathcal{G}_n'$ using Eq. (6.7)
  end for
end if
return $\Theta' = \{s', \mathcal{G}_t', \mathcal{G}_n'\}$

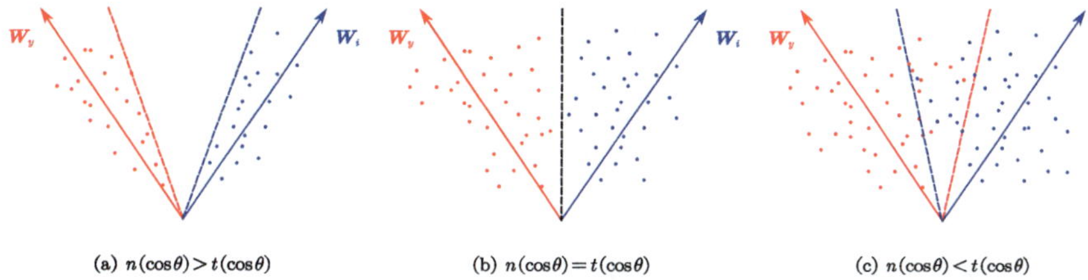

Fig. 6.4 Different margin under different settings. The dotted line represents the decision boundary of each class

1. $t'(x) \geq 0, x \in [-1, 1]$;
2. $n'(x) \geq 0, x \in [-1, 1]$.

According to the above two properties, the composite function $t(\cos\theta)$ and $n(\cos\theta)$ are generally decreasing functions with $\theta \in [0, \pi]$.

For clarity, we focus solely on the binary scenario within $\mathcal{L}_{gms}$ (where the target class is y and the non-target class is i) to demonstrate the third property. The decision boundary for the target class y is defined by $t(\cos\theta_y) - n(\cos\theta_i) = 0$. When $n(\cos\theta) = t(\cos\theta)$, the decision boundary corresponds to $\theta_y = \theta_i$, as depicted in Fig. 6.4b. When $n(\cos\theta) < t(\cos\theta)$, the decision boundary implies $\theta_y - \theta_i > 0$, leading to indistinguishable positive and negative samples within a certain region, as shown in Fig. 6.4c. When $n(\cos\theta) > t(\cos\theta)$, the decision boundary corresponds to $\theta_y - \theta_i < 0$, illustrated in Fig. 6.4a. This scenario creates a larger margin between positive and negative samples, facilitating the learning of more discriminative features. Thus, the third property

$$n(x) - t(x) \geq 0, \quad x \in [-1, 1]$$

is an essential characteristic of the GMS loss function.

Given the complexity of verifying whether the continuous functions $t(x)$ and $n(x)$ satisfy the aforementioned properties, we utilize f_t and f_n in Eq. (6.11) for testing purposes. To enhance the flexibility of our search space in practical implementation, we relax the requirements for these properties. Specifically, we stipulate that only τ_p percent of the values in f_t and f_n must meet the properties. This relaxation allows CircleLoss, which does not strictly adhere to the second property, to pass the protocol.

Toy task. We draw B random samples $\{(\boldsymbol{x}_b, y_b)\}|_{b=1}^{B}$ from the training dataset $\mathcal{D}_{trn}$ and initialize a network $\mathcal{M}_{\Omega_0}$ randomly. We then record the network's predictions and corresponding labels as $\{(\hat{\boldsymbol{x}}_b = \mathcal{M}_{\Omega_0}(\boldsymbol{x}_b), y_b)\}|_{b=1}^{B}$. The toy task is designed to optimize solely over $\{\hat{\boldsymbol{x}}_b\}|_{b=1}^{B}$ to solve the following problem:

$$\{\hat{\boldsymbol{x}}_b^*\}|_{b=1}^{B} = \underset{\{\hat{\boldsymbol{x}}_b\}|_{b=1}^{B}}{\arg\min} \frac{1}{B} \sum_{b=1}^{B} \mathcal{L}_{gms}^{\Theta}\left(\hat{\boldsymbol{x}}_b, y_b\right), \tag{6.10}$$

To summarize, the loss-rejection protocol initially evaluates the basic properties of Θ and subsequently assesses the toy task. Only those Θ that fulfill the basic properties and achieve an evaluation metric $\xi(\{\hat{\boldsymbol{x}}_b^*\}|_{b=1}^B)$ above the predefined threshold τ_{toy} on the toy task are permitted to pass; otherwise, they are rejected.

6.3.2.3 Equivalence-Check Strategy

While AutoLoss-Zero employs the initial gradient norm $\left\{\left\|\partial\mathcal{L}_{gms}^{\Theta}/\partial\hat{\boldsymbol{x}}_b\right\|_2\right\}|_{b=1}^B$ in the toy task to identify mathematically equivalent loss functions, this approach may be imprecise when relying solely on the initial gradient norm. By considering the interplay among s, $t(x)$, and $n(x)$ in $\mathcal{L}_{gms}$, we can formalize Θ as a concise feature vector that accurately identifies equivalent loss functions.

First, we discretize Θ as follows:

$$d_\Theta = \left[f_t, f_n, \log_2(s)/(\Delta_s N_s)\right] \in \mathbb{R}^{2N+1}, \tag{6.11}$$

where $f_t \in \mathbb{R}^N$ and $f_n \in \mathbb{R}^N$ represent the values of $t(x)$ and $n(x)$ at uniformly spaced points within the domain $[-1, 1]$. However, d_Θ cannot distinguish mathematically equivalent loss functions under translation-scale transformations. For instance, $\Theta_0 = \{t(x), n(x), s\}$ and $\Theta_{k,b} = \{t(x)/k + b, n(x)/k + b, ks\}$ (for all $k \neq 0$ and $b \in \mathbb{R}$) are mathematically equivalent for $\mathcal{L}_{gms}$. To address this, we introduce a normalization transformation for $t(x)$ and $n(x)$:

$$\begin{cases} TN_{\min} = \min\left\{\min\limits_{x\in[-1,1]} t(x), \min\limits_{x\in[-1,1]} n(x)\right\} \\ TN_{\max} = \max\left\{\max\limits_{x\in[-1,1]} t(x), \max\limits_{x\in[-1,1]} n(x)\right\} \\ b = (TN_{\min} + TN_{\max})/2 \\ k = \max\left\{(TN_{\max} - TN_{\min})/2, 1/s\right\} \\ \bar{s} = k \cdot s \\ \bar{t}(x) = (t(x) - b)/k \\ \bar{n}(x) = (n(x) - b)/k \end{cases}, \tag{6.12}$$

where the transformed $\bar{s} \geq 1$, and the range of $\bar{t}(x)$ and $\bar{n}(x)$ is within $[-1, 1]$. This normalized feature vector $d_{\bar{\Theta}}$ can accurately identify mathematically equivalent loss functions under translation-scale transformations. However, $d_{\bar{\Theta}}[2N] \in [0, +\infty]$, which significantly impacts the prediction performance of the predictor discussed in the next section. To mitigate this, we propose a scale constraint on the search space. Specifically, the generated computational graph (CG) must satisfy:

$$\log_2((TN_{\max} - TN_{\min}) \cdot s/2) \leq \Gamma, \tag{6.13}$$

where $\Gamma > \Delta_s N_s$. Under this constraint, the value range of $\log_2(\overline{s})$ is confined to $[0, \Gamma]$. In summary, we define the feature vector of a loss function parameterized by Θ as:

$$fv_\Theta = [f_{\overline{t}}, f_{\overline{n}}, 2\log_2(\overline{s})/\Gamma - 1]. \tag{6.14}$$

Each element in fv_Θ lies within the interval $[-1, 1]$, facilitating normalization for predictor training. Moreover, fv_Θ serves as an equivalence-check mechanism to identify mathematically equivalent loss functions, thereby avoiding redundant evaluations.

6.3.2.4 Promising-Loss Chooser

While the loss-rejection protocol effectively filters out invalid loss functions, some functions that pass the protocol still exhibit subpar performance. To conserve computational resources, we draw inspiration from performance predictors in neural architecture search (NAS) (Dudziak et al. 2020; Xu et al. 2021) and introduce a CNN-based performance predictor for loss functions. This predictor aims to identify the most promising-loss functions from the candidate pool.

Once the number of loss functions evaluated on the proxy task reaches E_0, the predictor $\mathcal{P}$ is trained on the current set of evaluated pairs (Θ_i, p_i), denoted as $Eva = \{(\Theta_i, p_i)\}|_{i=1}^{E_0}$. Subsequently, every time the size of Eva increases by ΔE, the predictor $\mathcal{P}$ is updated based on the latest Eva set. After the initial training of the predictor, each newly generated loss function that passes the equivalence-check is added to the candidate pool of the promising-loss chooser. When the candidate pool reaches the predefined size N_p, the predictor selects the most promising loss function, and the pool is then cleared. The detailed algorithm for selecting the promising loss function is outlined in Algorithm 6.4.

In this chapter, we select ResNet (He et al. 2016) as the basis for our CNN-based predictor. The commonly used ResNet50 is selected as our predictor, and it needs to be adjusted according to our problem:

- The original ResNet is for two-dimensional data, but our data is one-dimensional. So the two-dimensional operations (Conv2d, BatchNorm2d, Pool2d) in the original ResNet are replaced with the one-dimensional operations (Conv1d, BatchNorm1d, Pool1d) to obtain 1D-ResNet.
- The input feature vector fv_Θ is reshaped to $x_\Theta \in \mathbb{R}^{3\times N}$ to have a compact representation of the loss function. The conversion from fv_Θ to x_Θ is as follows:

$$\begin{cases} x_\Theta[0,:] = fv_\Theta[:N] \\ x_\Theta[1,:] = fv_\Theta[N:2N] \\ x_\Theta[2,:] = fv_\Theta[2N] \cdot \mathbb{1} \end{cases}$$

- The feature after the last AvgPool1d in 1D-ResNet is passed through the fully connected layers to get the final predicted performance.

Algorithm 6.4 Promising-Loss Chooser

Require: A loss function Θ_i, the current evaluated set $Eva = \{(\Theta_e, p_e)\}|_{e=1}^{E}$, the predictor $\mathcal{P}$, the size of evaluated set to start predictor E_0, the interval for updating predictor ΔE, the current candidate population L, the maximum size of candidate population N_p.

Ensure: The loss function Θ_o to be evaluated on the proxy task.

if $E < E_0$ **then** ▷ Predictor doesn't work.
  $\Theta_o \leftarrow \Theta_i$
else if $E = E_0$ **then** ▷ Initialize predictor.
  Append Θ_i to L
  Initialize $\mathcal{P}$
  Update $\mathcal{P}$ with Eva
  return None
else ▷ Predictor works.
  Append Θ_i to L
  if $(E - E_0) \pmod{\Delta E} = 0$ **then**
    Update $\mathcal{P}$ with Eva
  end if
  if $|L| = N_p$ **then**
    $L_p \leftarrow [\,]$
    for each Θ_e in L **do**
      Append $\mathcal{P}(\Theta_e)$ to L_p
    end for
    $e_o \leftarrow \underset{e=0,1,\ldots,N_p-1}{\arg\max} \; L_p[e]$
    $\Theta_o \leftarrow L[e_o]$
    $L \leftarrow [\,]$
  else
    return None
  end if
end if
return Θ_o

Our predictor is trained with SGD optimizer, and the learning rate is fixed as 0.0035. The batch size is set to 16, and the τ in $\mathcal{L}_K$ is set to 0.01. The λ is set to 1.0, and a total of 200 epochs is used to train the predictor.

Given B distinct converted loss functions and their corresponding ground-truth performances $\{(x_{\Theta_i}, y_i)\}|_{i=1}^{B}$, the predictor outputs $\{\mathcal{P}(x_{\Theta_i})\}|_{i=1}^{B}$. The commonly used mean squared error (MSE) loss function is defined as:

$$\mathcal{L}_{MSE} = \frac{1}{B}\sum_{i=1}^{B}(\mathcal{P}(x_{\Theta_i}) - y_i)^2. \tag{6.15}$$

As noted in Dudziak et al. (2020) and Xu et al. (2021), a predictor that determines the ranking of loss functions is more robust than one that precisely predicts their performance. Kendall's Tau (KTau) (Sen et al. 1968) serves as an indicator to measure the ranking relationship:

$$\text{KTau} = \frac{1}{C_B^2} \sum_{1 \le i < j \le B} \text{sign}(y_i - y_j) \cdot \text{sign}(\mathcal{P}(x_{\Theta_i}) - \mathcal{P}(x_{\Theta_j})), \tag{6.16}$$

where $\text{sign}(x)$ is the sign function. Inspired by Brown et al. (2020), we convert the non-differentiable KTau to a differentiable version by replacing the second $\text{sign}(x)$ in Eq. (6.16) with $\tanh(x/\tau)$:

$$\mathcal{L}_K = \frac{1}{C_B^2} \sum_{1 \le i < j \le B} \text{sign}(y_i - y_j) \cdot \tanh\left(\frac{\mathcal{P}(x_{\Theta_i}) - \mathcal{P}(x_{\Theta_j})}{\tau}\right), \tag{6.17}$$

where τ controls the"temperature" of the $\tanh(x)$ function, effectively replacing the $\text{sign}(x)$ function. Consequently, the final loss function for our predictor is:

$$\mathcal{L} = \mathcal{L}_{MSE} + \lambda \mathcal{L}_K, \tag{6.18}$$

where λ is a hyperparameter that balances the importance of the two loss components.

6.4 Experiments

6.4.1 Datasets and Evaluation Metrics

To evaluate the performance of AutoLoss-GMS, we conducted experiments on three widely used datasets: Market-1501 (Zheng et al. 2015), CUHK03 (Ristani et al. 2016), and MSMT17 (Wei et al. 2018) (dataset statistics are provided in Table 6.3).

Adhering to the conventions of the ReID community (Zheng et al. 2015; Wei et al. 2018), we assessed all methods using Cumulative Matching Characteristic (CMC) curves and mean Average Precision (mAP).

Table 6.3 Dataset statistics. T: Train. Q: Query. G: Gallery

Dataset	# IDs (T-Q-G)	# images (T-Q-G)
Market-1501	751-750-751	12936-3368-15913
CUHK03	767-700-700	7365-1400-5332
MSMT17	1041-3060-3060	32621-11659-82161

Table 6.4 The searched loss functions. PIP: predefined initial populations. Each loss is searched on a certain dataset using certain model with PIP or not

Searched Loss	Model	Dataset	PIP	$t(x)$	$n(x)$	$log_2(s)$
AutoLoss-GMS-Zero	ResNet50	Market-1501	✗	$(0.22 + e^{\sqrt{0.22}})x + \arcsin(0.22)^2$	$x + 0.85$	5.5
AutoLoss-GMS-A	ResNet50	Market-1501	✓	$\text{de}(1 - x)(x - 0.7)$	$[\text{Gd}(x - \text{Sig}(0.25))]_+$	7.5
AutoLoss-GMS-B	ResNet50	CUHK03	✓	$\text{de}(1.3 - x)(x - 1.0)$	$0.35x - 0.35^2$	6.0
AutoLoss-GMS-C	OSNet	Market-1501	✓	$(x - 0.84)(0.95 - x)$	$[\text{de}(\arcsin(x))]_+(x - 0.5) + 0.05$	7.5
AutoLoss-GMS-D	MGN	Market-1501	✓	$x + 0.15$	$x + 0.2$	4.0

6.4.2 Implementation Details

6.4.2.1 Search Configurations

The search process is conducted on five NVIDIA Tesla-V100 GPUs and takes approximately 2.5 days to complete. Under various configurations, five distinct loss functions are identified, the details of which are presented in Table 6.4.

Search algorithm. To kick off the search procedure, a population comprising $K = 20$ loss functions is established, with subsequent iterations retaining merely the latest $P = 1000$ loss functions. In the scenario of de novo search, the entire cohort of $K = 20$ loss functions at genesis is conjured up via random generation, in accordance with Algorithm 6.2. When leveraging prior knowledge, a blend of ten randomly generated loss functions and another ten loss functions, which are instantiated based on hand-crafted ones (their hyperparameters are configured in line with the grid search outcomes depicted in Fig. 6.5), is employed. The proportion for tournament selection is calibrated to $T = 5\%$ of the extant population. The search expedition comes to a halt once 500 models have been scrutinized on the proxy task. A comprehensive roster of additional hyperparameters deployed in AutoLoss-GMS is cataloged in Table 6.5.

Proxy task. For each dataset, we designate 80% of the pedestrian images from the original training set as the search training dataset, with the remainder serving as the validation dataset. To enhance search efficiency, we halve the image resolution in the proxy task compared to the actual training task, while maintaining other configurations consistent with the training settings.

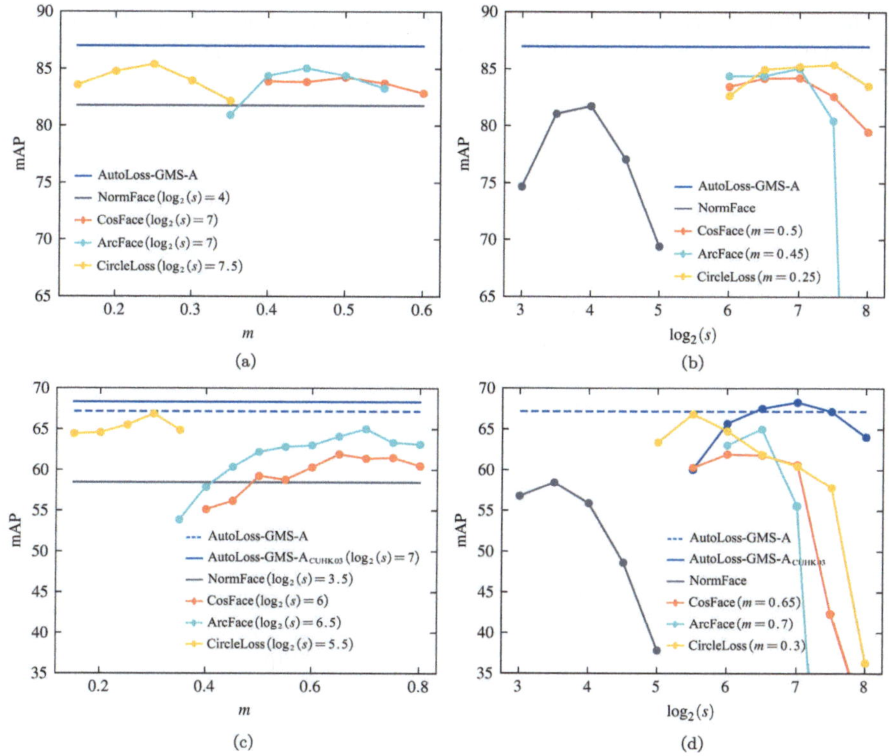

Fig. 6.5 Comparison of hyperparameter impact between hand-crafted loss functions and the searched AutoLoss-GMS-A. Figures (**a**) and (**b**) show the results for the ResNet50 model on the Market-1501 dataset. Figures (**c**) and (**d**) display the corresponding results for the ResNet50 model on the CUHK03 dataset. These plots demonstrate how performance varies with different hyperparameter settings for both types of loss functions

6.4.2.2 Train Configurations

The loss function identified as optimal during the search phase is retrained on the complete original training dataset.

Data preprocessing. For ResNet50 and OSNet, images are resized to 256×128. For MGN, we adhere to the original paper's specifications, using an image size of 384×128. Data augmentation techniques include random cropping, random horizontal flipping, and random erasing.

Data sampling. We implement the PK sampling strategy with $P = 16$ identities and $K = 4$ samples per identity.

Training. Utilizing the SGD optimizer with a momentum of 0.9 and weight decay of 5×10^{-4}, we train the models for a total of 140 epochs. A warmup learning rate strategy is employed, where the first 20 epochs involve linearly increasing the learning rate from 0.0001 to 0.01, with the backbone weights held constant. Subsequently,

Table 6.5 The hyperparameters used in AutoLoss-GMS. CG: computational graph. POS: produce the offspring. LRP: loss-rejection protocol. ECS: equivalence-check strategy. PLC: promising-loss chooser

CG	n_c	2
	n_o	10
	Δs	0.5
	N_s	20
	Δc	0.05
	N_c	20
POS	p_c	0.6
	p_{copy}	0.1
	p_{re}	0.4
	N_m	3
LRP	τ_p	80
	τ_{toy}	0.9
	B	64
ECS	N	257
	Γ	13
PLC	E_0	100
	ΔE	100
	N_p	10

the learning rate is reduced to 0.001 at the 60th epoch and to 0.0001 at the 100th epoch.

Testing. During testing, test-time flipping is applied, and cosine similarity is used as the evaluation metric.

6.4.3 Ablation Study

6.4.3.1 Analysis of the Predictor

To assess the predictor's components, we initially sample 1000 random loss functions and evaluate their performance on the proxy task. Across various configurations, we conduct ten training iterations, each time randomly partitioning the data into training and test sets. The outcomes are depicted in Fig. 6.6a, yielding the following insights:

- The predictor trained solely with $\mathcal{L}_{MSE}$ underperforms compared to the combination $\mathcal{L}_{MSE} + \mathcal{L}_K$, underscoring the efficacy of our differentiable KTau.
- The unnormalized d_Θ yields suboptimal results.
- Transforming $f v_\Theta$ into a three-channel x_Θ enhances performance.

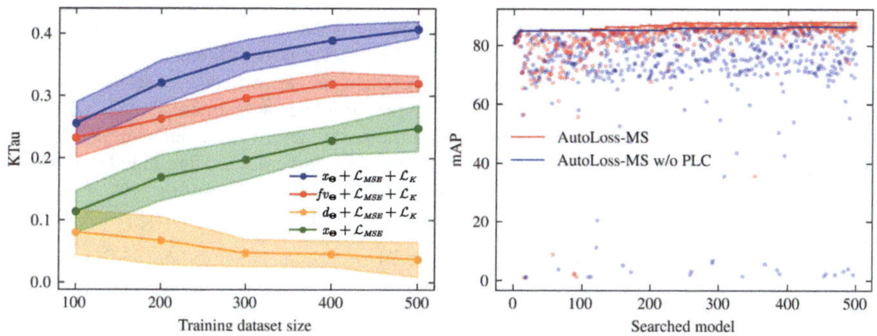

(a) (b)

Fig. 6.6 The analysis of the predictor. PLC: promising-loss chooser

Table 6.6 Component analysis of AutoLoss-GMS for searching on Market-1501 with ResNet50. This figure illustrates the contribution of each proposed component to the overall performance and efficiency of the search process. The components evaluated are: Random Search (RS), Evolutionary Algorithm (EA), Cross-Graph Mutation (CGM), Loss-Rejection Protocol (LRP), Basic Properties Protocol (BPP), Toy Task (TT), Equivalence-Check Strategy (ECS), Predefined Initial Population (PIP), and Promising-Loss Chooser (PLC)

		mAP	Speed-up	Explored losses
RS	Backward	81.32	1×	~ 500
	Forward	82.67	1×	~ 500
	Naïve EA	83.39	1×	~ 500
	+CGM	83.77	1×	~ 500
	+LRP(BPP)	84.52	~ 30×	~ 1.5× 10^4
	+LRP(TT)	84.95	~ 36×	~ 1.8× 10^4
	+ECS	85.24	~ 54×	~ 2.7× 10^4
	+PIP	85.80	~ 58×	~ 2.9× 10^4
	+PLC	**88.02**	~ **540**×	~ **2.7**× $\mathbf{10^5}$

In Fig. 6.6b, we compare the performance of AutoLoss-GMS both with and without the predictor. The results indicate that once the predictor is introduced after training 100 loss functions, a significant number of poorly performing loss functions are filtered out. This further underscores the effectiveness of our promising-loss chooser.

6.4.3.2 Effectiveness of the Components in AutoLoss-GMS

To assess the impact of various components in AutoLoss-GMS, we conducted searches with different configurations, and the results are presented in Table 6.6. From these results, we can draw the following conclusions:

Table 6.7 Comparison of transferability for searched and hand-crafted losses, reporting only the mean Average Precision (mAP) metric. The results for ResNet50 and OSNet are denoted by "R" and "O", respectively. Values inside parentheses represent the performance of losses transferred directly across datasets or networks without any hyperparameter adjustment. Values outside the parentheses show the performance after fine-tuning the scale factor (s) for our searched losses or other hyperparameters for hand-crafted losses. A single value indicates the fine-tuned performance for hand-crafted losses or the loss performance found under that specific setting

Methods		Market-1501	CUHK03
R	NormFace	81.74	58.48(55.91)
	CosFace	84.23	61.94(57.41)
	ArcFace	85.06	65.03(56.54)
	CircleLoss	85.41	66.87(59.21)
	AutoLoss-GMS-A	**87.00**	**68.33**(67.18)
	AutoLoss-GMS-B	78.03(76.10)	67.11
	AutoLoss-GMS-C	86.64(86.23)	67.95(67.79)
	AutoLoss-GMS-Zero	84.16	64.32(64.32)
O	AutoLoss-GMS-A	**88.99**(88.87)	70.60(70.60)
	AutoLoss-GMS-B	86.20(84.65)	71.15(70.27)
	AutoLoss-GMS-C	88.94	**72.57**(72.17)
	AutoLoss-GMS-Zero	85.79(84.59)	68.95(67.59)

- The forward-construction method outperforms the backward construction method, confirming that the computational graphs (CGs) constructed via the forward method yield a more diverse set of loss functions.
- The proposed cross-graph mutation operation enhances search performance.
- The two-step loss-rejection protocol expands the search space by 36 times and rapidly filters out loss functions that fail to meet basic properties or perform poorly on the toy task.
- The equivalence-check strategy enables the exploration of 50% more loss functions by avoiding redundant evaluations of equivalent loss functions.
- Incorporating prior knowledge through hand-crafted loss functions improves search outcomes.
- The predictor-based promising-loss chooser boosts search efficiency by approximately 10 times.

6.4.3.3 Transferability of the Searched Losses

To evaluate the transferability of the loss functions discovered on various datasets and networks, we conducted comprehensive comparisons, as detailed in Table 6.7. The hyperparameters of the hand-crafted loss functions were meticulously fine-tuned on the two datasets via grid search, as shown in Fig. 6.5. Our searched AutoLoss-GMS-A achieved a mean Average Precision (mAP) that is 1.59% higher than the best-performing CircleLoss among the hand-crafted loss functions. Without a predefined initial population, AutoLoss-GMS-Zero can match the performance of CosFace. When the hyperparameters initially set for Market-1501 are applied, AutoLoss-GMS-A can achieve an mAP of 67.18% on the CUHK03 dataset, surpassing CircleLoss's 59.21%. Further fine-tuning of CircleLoss on CUHK03 can boost its performance to 66.87%. To rapidly adapt to other datasets or networks, we can fine-tune the scale parameter s of the searched loss. AutoLoss-GMS-A, after fine-tuning on CUHK03, reaches an mAP of 68.33%, which is 1.46% higher than CircleLoss.

AutoLoss-GMS-B and AutoLoss-GMS-C were tested for their performance across datasets and networks alongside AutoLoss-GMS-A. AutoLoss-GMS-B performs well on CUHK03 but not as effectively on Market-1501. This discrepancy is primarily due to CUHK03 having less data than Market-1501. Thus, searching on a larger dataset yields a loss function with better generalization capabilities. AutoLoss-GMS-A and AutoLoss-GMS-C were obtained by searching on different models. Irrespective of fine-tuning on s, these two loss functions consistently deliver top-tier performance under their respective models. Even when transferred, the performance of the searched loss functions is sufficient to outperform hand-crafted loss functions, demonstrating a certain degree of transferability.

6.4.4 The Analysis of the Searched Loss Functions

The specific expressions of the searched loss functions are shown in Table 6.4, and the $t(x)$ and $n(x)$ of the hand-crafted loss functions and the searched loss functions are shown in Fig. 6.7. First of all, the $t(x)$ and $n(x)$ in AutoLoss-GMS-Zero and AutoLoss-GMS-D are both linear functions. Due to the lack of prior knowledge, AutoLoss-GMS-Zero will fall into this simple form of a local solution. For AutoLoss-GMS-D, because the triplet loss function is involved in the search process, it is possible that this simple form can bring good results.

The forms of AutoLoss-GMS-A, AutoLoss-GMS-B and AutoLoss-GMS-C are similar to CircleLoss, but there are still some differences. Specifically, none of the three searched $t(x)$ and $n(x)$ have the Detach operation simultaneously. Although the $t(x)$ of the three searched loss functions and the $t(x)$ of CircleLoss are similar to quadratic functions, their $n(x)$s are far from CircleLoss's, which are difficult to be designed by hand.

6.4.5 Comparison With State-of-the-Art Methods

Our method is compared against state-of-the-art (SOTA) approaches using two distinct types of models: those based on global features and those based on local features. We report the performance of three variants of our method: AutoLoss-GMS-A, AutoLoss-GMS-C, and AutoLoss-GMS-D. The results of this comparison are detailed in Table 6.8.

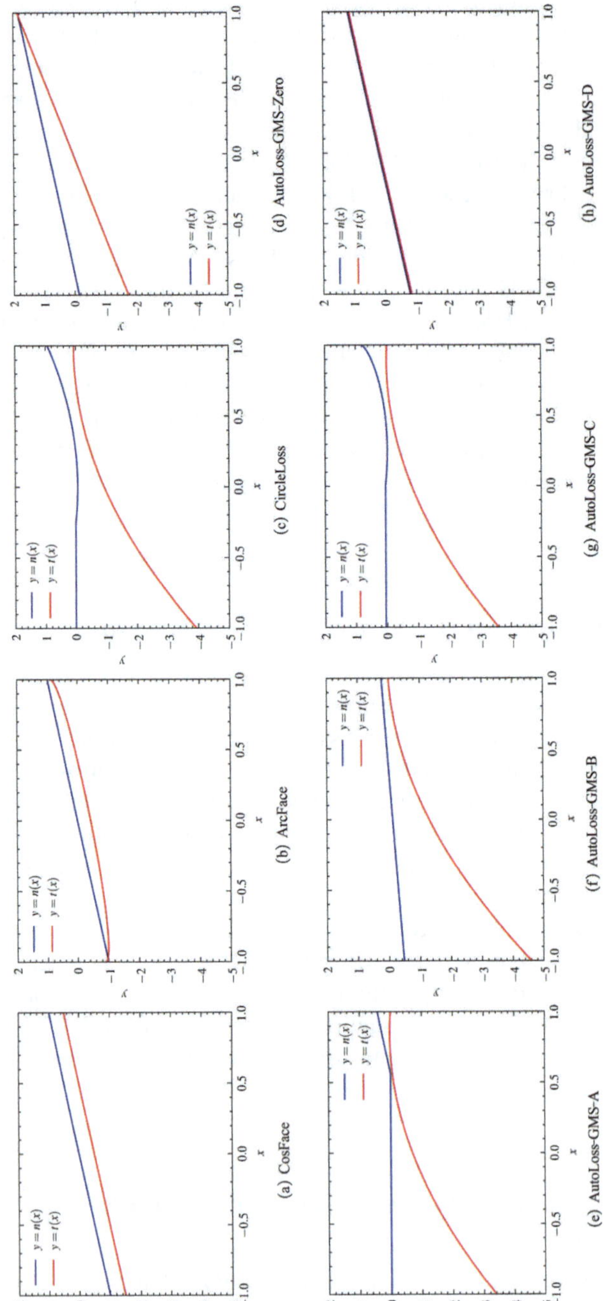

Fig. 6.7 The $t(x)$ and $n(x)$ of the hand-crafted loss functions and the searched loss functions

Table 6.8 Comparison with state-of-the-art methods. The table is divided into two sections by the double line: the top section for models based on global features and the bottom for models based on local features. In each section, our methods are listed below the single line, while other methods are above. The number in parentheses after a loss function indicates how many times it was used. The **red bold** font highlights the best performance achieved by our methods, and the ***blue bold*** font denotes the best performance among other methods

Methods	Market-1501		CUHK03		MSMT17	
	mAP	Rank-1	mAP	Rank-1	mAP	Rank-1
TriNet (Hermans et al. 2017)	69.1	84.9	–	–	–	–
SphereReID (Fan et al. 2019)	83.6	94.4	-0-	–	–	–
StrongBaseline (Luo et al. 2019)	85.9	94.5	–	–	–	–
ResNet50+CircleLoss (Sun et al. 2020)	84.9	94.2	–	–	50.2	76.3
OSNet (Zhou et al. 2019)	*86.7*	94.8	67.8	72.3	*55.1*	*79.1*
AutoReID (Quan et al. 2019)	85.1	94.5	*69.3*	*73.3*	52.5	78.2
SphereReID+AM-LFS (Li et al. 2019)	84.4	*95.0*	–	–	–	–
ResNet50+AutoLoss-GMS-A(ours)	87.0	94.7	68.3	70.4	55.1	79.5
OSNet+AutoLoss-GMS-C(ours)	**88.9**	**95.7**	**72.6**	**74.5**	**62.6**	**83.7**
PCB+RPP (Sun et al. 2018)	81.6	93.8	57.5	63.7	–	–
MGN (Wang et al. 2018a)	86.9	95.7	*66.0*	*66.8*	–	–
MGN+CircleLoss (Sun et al. 2020)	87.4	*96.1*	–	–	*52.1*	*76.9*
MGN+AM-LFS (Li et al. 2019)	*88.1*	95.8	–	–	–	–
MGN+AutoLoss-GMS-A(ours)	88.7	95.6	73.2	75.2	58.2	80.9
MGN+AutoLoss-GMS-C(ours)	89.3	95.7	72.3	75.3	58.3	79.9
MGN+AutoLoss-GMS-D(ours)	**90.1**	**96.2**	**74.3**	**75.6**	**63.0**	**83.7**

Among the models based on the global feature with the ResNet50 backbone, ResNet50 + AutoLoss-GMS-A achieves the highest mAP on all three datasets. Especially on Market-1501, our AutoLoss-GMS-A can surpass StrongBaseline, which uses cross-entropy and triplet paradigms, by 1.1%. On the largest dataset, MSMT17, our AutoLoss-GMS-A significantly surpasses CircleLoss by 4.9% in terms of mAP. On all three datasets, under the support of AutoLoss-GMS-A, ResNet50 can achieve similar performance with the network architecture specially designed for ReID (i.e., OSNet, AutoReID). When equipped with OSNet, AutoLoss-GMS-C can further improve the original OSNet by a significant step. Especially on MSMT17, OSNet + AutoLoss-GMS-C can reach 62.6% mAP.

The performance of directly applying AutoLoss-GMS-A and AutoLoss-GMS-C to the MGN model is also significantly better than MGN + CircleLoss, which further verifies that our searched loss functions are transferable across models. Furthermore, MGN + AutoLoss-GMS-D surpasses other methods on Market-1501 with mAP of 90.1%, respectively.

In summary, the performance of our searched loss functions is significantly better than the performance of the hand-crafted CircleLoss and AutoML-based AM-LFS, which further illustrates the effectiveness of our method.

6.5 Conclusion

This chapter presents AutoLoss-GMS, a novel method for automatically discovering superior loss functions within the generalized margin-based softmax (GMS) loss space for person re-identification. Our experimental results show that the loss functions found by our method achieve state-of-the-art performance. A key finding is that these searched loss functions are transferable, meaning they can be applied to different models and datasets without significant performance degradation, a crucial property for practical applications.

Despite these successes, our current best search results still rely heavily on the inclusion of prior knowledge (e.g., initializing the search with known, high-performing loss functions). Future research will focus on enhancing our algorithm to achieve excellent performance even when starting the search from scratch, without any predefined knowledge. Additionally, we plan to apply our search framework to a wider variety of tasks to further validate the versatility and effectiveness of our proposed algorithm.

References

Alexander Hermans, Lucas Beyer, and Bastian Leibe. In defense of the triplet loss for person re-identification. *CoRR*, abs/1703.07737, 2017.

Andrew Brown, Weidi Xie, Vicky Kalogeiton, and Andrew Zisserman. Smooth-ap: Smoothing the path towards large-scale image retrieval. In *European Conference on Computer Vision*, pages 677–694. Springer, 2020.

Barret Zoph and Quoc V. Le. Neural architecture search with reinforcement learning. In *5th International Conference on Learning Representations, ICLR 2017, Toulon, France, April 24–26, 2017, Conference Track Proceedings*. OpenReview.net, 2017.

Chuming Li, Xin Yuan, Chen Lin, Minghao Guo, Wei Wu, Junjie Yan, and Wanli Ouyang. AM-LFS: automl for loss function search. In *2019 IEEE/CVF International Conference on Computer Vision, ICCV 2019, Seoul, Korea (South), October 27-November 2, 2019*, pages 8409–8418. IEEE, 2019.

Ekin D Cubuk, Barret Zoph, Dandelion Mane, Vijay Vasudevan, and Quoc V Le. Autoaugment: Learning augmentation strategies from data. In *Proceedings of the IEEE/CVF Conference on Computer Vision and Pattern Recognition*, pages 113–123, 2019.

Ergys Ristani, Francesco Solera, Roger Zou, Rita Cucchiara, and Carlo Tomasi. Performance measures and a data set for multi-target, multi-camera tracking. In *European conference on computer vision*, pages 17–35. Springer, 2016.

Esteban Real, Alok Aggarwal, Yanping Huang, and Quoc V Le. Regularized evolution for image classifier architecture search. In *Proceedings of the aaai conference on artificial intelligence*, volume 33, pages 4780–4789, 2019.

Esteban Real, Chen Liang, David So, and Quoc Le. Automl-zero: Evolving machine learning algorithms from scratch. In *International Conference on Machine Learning*, pages 8007–8019. PMLR, 2020.

Xing Fan, Wei Jiang, Hao Luo, and Mengjuan Fei. Sphereid: Deep hypersphere manifold embedding for person re-identification. *Journal of Visual Communication and Image Representation*, 60:51–58, 2019.

Feng Wang, Xiang Xiang, Jian Cheng, and Alan Loddon Yuille. Normface: L_2 hypersphere embedding for face verification. In Qiong Liu, Rainer Lienhart, Haohong Wang, Sheng-Wei "Kuan-Ta" Chen, Susanne Boll, Yi-Ping Phoebe Chen, Gerald Friedland, Jia Li, and Shuicheng Yan, editors, *Proceedings of the 2017 ACM on Multimedia Conference, MM 2017, Mountain View, CA, USA, October 23–27, 2017*, pages 1041–1049. ACM, 2017.

Feng Zheng, Cheng Deng, Xing Sun, Xinyang Jiang, Xiaowei Guo, Zongqiao Yu, Feiyue Huang, and Rongrong Ji. Pyramidal person re-identification via multi-loss dynamic training. In *Proceedings of the IEEE/CVF Conference on Computer Vision and Pattern Recognition*, pages 8514–8522, 2019.

Guangyi Chen, Tianpei Gu, Jiwen Lu, Jin-An Bao, and Jie Zhou. Person re-identification via attention pyramid. *IEEE Transactions on Image Processing*, 2021.

Guanshuo Wang, Yufeng Yuan, Xiong Chen, Jiwei Li, and Xi Zhou. Learning discriminative features with multiple granularities for person re-identification. In *Proceedings of the 26th ACM international conference on Multimedia*, pages 274–282, 2018.

Hanxiao Liu, Karen Simonyan, and Yiming Yang. DARTS: differentiable architecture search. In *7th International Conference on Learning Representations, ICLR 2019, New Orleans, LA, USA, May 6–9, 2019*. OpenReview.net, 2019.

Hao Li, Chenxin Tao, Xizhou Zhu, Xiaogang Wang, Gao Huang, and Jifeng Dai. Auto seg-loss: Searching metric surrogates for semantic segmentation. In *9th International Conference on Learning Representations, ICLR 2021, Virtual Event, Austria, May 3–7, 2021*. OpenReview.net, 2021.

Hao Li, Tianwen Fu, Jifeng Dai, Hongsheng Li, Gao Huang, and Xizhou Zhu. Autoloss-zero: Searching loss functions from scratch for generic tasks. *CoRR*, abs/2103.14026, 2021.

Hao Luo, Youzhi Gu, Xingyu Liao, Shenqi Lai, and Wei Jiang. Bag of tricks and a strong baseline for deep person re-identification. In *Proceedings of the IEEE Conference on Computer Vision and Pattern Recognition Workshops*, pages 0–0, 2019.

Hao Wang, Yitong Wang, Zheng Zhou, Xing Ji, Dihong Gong, Jingchao Zhou, Zhifeng Li, and Wei Liu. Cosface: Large margin cosine loss for deep face recognition. In *Proceedings of the IEEE conference on computer vision and pattern recognition*, pages 5265–5274, 2018.

Hongyang Gu, Guangyuan Fu, Jianmin Li, and Jun Zhu. Auto-reid+: Searching for a multi-branch convnet for person re-identification. *Neurocomputing*, 435:53–66, 2021.

Jiankang Deng, Jia Guo, Niannan Xue, and Stefanos Zafeiriou. Arcface: Additive angular margin loss for deep face recognition. In *Proceedings of the IEEE/CVF Conference on Computer Vision and Pattern Recognition*, pages 4690–4699, 2019.

Kaiming He, Xiangyu Zhang, Shaoqing Ren, and Jian Sun. Deep residual learning for image recognition. In *Proceedings of the IEEE conference on computer vision and pattern recognition*, pages 770–778, 2016.

Kaiyang Zhou, Yongxin Yang, Andrea Cavallaro, and Tao Xiang. Omni-scale feature learning for person re-identification. In *Proceedings of the IEEE International Conference on Computer Vision*, pages 3702–3712, 2019.

Keyu Tian, Chen Lin, Ming Sun, Luping Zhou, Junjie Yan, and Wanli Ouyang. Improving auto-augment via augmentation-wise weight sharing. In Hugo Larochelle, Marc'Aurelio Ranzato, Raia Hadsell, Maria-Florina Balcan, and Hsuan-Tien Lin, editors, *Advances in Neural Information Processing Systems 33: Annual Conference on Neural Information Processing Systems 2020, NeurIPS 2020, December 6–12, 2020, virtual*, 2020.

Liang Zheng, Liyue Shen, Lu Tian, Shengjin Wang, Jingdong Wang, and Qi Tian. Scalable person re-identification: A benchmark. In *Proceedings of the IEEE international conference on computer vision*, pages 1116–1124, 2015.

Longhui Wei, Shiliang Zhang, Wen Gao, and Qi Tian. Person transfer gan to bridge domain gap for person re-identification. In *Proceedings of the IEEE conference on computer vision and pattern recognition*, pages 79–88, 2018.

Lukasz Dudziak, Thomas C. P. Chau, Mohamed S. Abdelfattah, Royson Lee, Hyeji Kim, and Nicholas D. Lane. BRP-NAS: prediction-based NAS using gcns. In Hugo Larochelle, Marc'Aurelio Ranzato, Raia Hadsell, Maria-Florina Balcan, and Hsuan-Tien Lin, editors, *Advances in Neural Information Processing Systems 33: Annual Conference on Neural Information Processing Systems 2020, NeurIPS 2020, December 6–12, 2020, virtual*, 2020.

Mang Ye, Jianbing Shen, Gaojie Lin, Tao Xiang, Ling Shao, and Steven CH Hoi. Deep learning for person re-identification: A survey and outlook. *IEEE Transactions on Pattern Analysis and Machine Intelligence*, 2021.

Max Jaderberg, Valentin Dalibard, Simon Osindero, Wojciech M. Czarnecki, Jeff Donahue, Ali Razavi, Oriol Vinyals, Tim Green, Iain Dunning, Karen Simonyan, Chrisantha Fernando, and Koray Kavukcuoglu. Population based training of neural networks. *CoRR*, abs/1711.09846, 2017.

Peidong Liu, Gengwei Zhang, Bochao Wang, Hang Xu, Xiaodan Liang, Yong Jiang, and Zhenguo Li. Loss function discovery for object detection via convergence-simulation driven search. In *9th International Conference on Learning Representations, ICLR 2021, Virtual Event, Austria, May 3–7, 2021*. OpenReview.net, 2021.

Pranab Kumar Sen. Estimates of the regression coefficient based on kendall's tau. *Journal of the American statistical association*, 63(324):1379–1389, 1968.

Ruijie Quan, Xuanyi Dong, Yu Wu, Linchao Zhu, and Yi Yang. Auto-reid: Searching for a part-aware convnet for person re-identification. In *Proceedings of the IEEE International Conference on Computer Vision*, pages 3750–3759, 2019.

Seyedali Mirjalili. Genetic algorithm. In *Evolutionary algorithms and neural networks*, pages 43–55. Springer, 2019.

Shuting He, Hao Luo, Pichao Wang, Fan Wang, Hao Li, and Wei Jiang. Transreid: Transformer-based object re-identification. *CoRR*, abs/2102.04378, 2021.

Sungbin Lim, Ildoo Kim, Taesup Kim, Chiheon Kim, and Sungwoong Kim. Fast autoaugment. In Hanna M. Wallach, Hugo Larochelle, Alina Beygelzimer, Florence d'Alché-Buc, Emily B. Fox, and Roman Garnett, editors, *Advances in Neural Information Processing Systems 32: Annual Conference on Neural Information Processing Systems 2019, NeurIPS 2019, December 8–14, 2019, Vancouver, BC, Canada*, pages 6662–6672, 2019.

Tianlong Chen, Shaojin Ding, Jingyi Xie, Ye Yuan, Wuyang Chen, Yang Yang, Zhou Ren, and Zhangyang Wang. Abd-net: Attentive but diverse person re-identification. In *Proceedings of the IEEE International Conference on Computer Vision*, pages 8351–8361, 2019.

Weiyang Liu, Yandong Wen, Zhiding Yu, Ming Li, Bhiksha Raj, and Le Song. Sphereface: Deep hypersphere embedding for face recognition. In *Proceedings of the IEEE conference on computer vision and pattern recognition*, pages 212–220, 2017.

Xiaobo Wang, Shuo Wang, Cheng Chi, Shifeng Zhang, and Tao Mei. Loss function search for face recognition. In *International Conference on Machine Learning*, pages 10029–10038. PMLR, 2020.

Xiaobo Wang, Shuo Wang, Cheng Chi, Shifeng Zhang, and Tao Mei. Loss function search for face recognition. In *Proceedings of the 37th International Conference on Machine Learning, ICML 2020, 13–18 July 2020, Virtual Event*, volume 119 of *Proceedings of Machine Learning Research*, pages 10029–10038. PMLR, 2020.

Xinyu Zhang, Qiang Wang, Jian Zhang, and Zhao Zhong. Adversarial autoaugment. In *8th International Conference on Learning Representations, ICLR 2020, Addis Ababa, Ethiopia, April 26–30, 2020*. OpenReview.net, 2020.

Yifan Sun, Changmao Cheng, Yuhan Zhang, Chi Zhang, Liang Zheng, Zhongdao Wang, and Yichen Wei. Circle loss: A unified perspective of pair similarity optimization. In *Proceedings of the IEEE/CVF Conference on Computer Vision and Pattern Recognition*, pages 6398–6407, 2020.

Yifan Sun, Liang Zheng, Yi Yang, Qi Tian, and Shengjin Wang. Beyond part models: Person retrieval with refined part pooling (and a strong convolutional baseline). In *Proceedings of the European Conference on Computer Vision (ECCV)*, pages 480–496, 2018.

Yixing Xu, Yunhe Wang, Kai Han, Yehui Tang, Shangling Jui, Chunjing Xu, and Chang Xu. Renas: Relativistic evaluation of neural architecture search. In *IEEE Conference on Computer Vision and Pattern Recognition, CVPR 2021, virtual, June 19–25, 2021*, pages 4411–4420. Computer Vision Foundation/IEEE, 2021.

Xinbang Zhang, Zehao Huang, Naiyan Wang, Shiming Xiang, and Chunhong Pan. You only search once: Single shot neural architecture search via direct sparse optimization. *IEEE Transactions on Pattern Analysis and Machine Intelligence*, 43(9):2891–2904, 2021.

Chapter 7
Conclusion and Future Work

7.1 Conclusion

With the continued development of global economies and societies, a new trend is emerging: the application of intelligent technologies, with urban intelligence at the forefront. Consequently, person re-identification, as a fundamental technology, has attracted increasing attention from researchers. Currently, deep learning-based person re-identification methods have achieved promising performance. However, their capabilities in many aspects still appear suboptimal, making it difficult to meet the practical application demands of complex, large-scale scenarios in current video surveillance. This book integrates machine learning techniques and, from three perspectives—data augmentation, network architecture, and loss functions—which are central to current models, proposes five reasonable and effective deep learning methods. These methods further enhance the accuracy, practicality, and robustness of person re-identification models. The main research content and contributions of this book are summarized as follows:

1. **Regarding the data augmentation strategy in deep learning-based person re-identification**: We proposed a data augmentation method based on erasable operations—*AutoErasing*. This data augmentation strategy transforms the non-differentiable erasing process into a differentiable one, building upon a differentiable erasing operation designed in this work. Besides searching for the size and position of the erased region produced by the erasing operation, it also considers searching for the color of the erased region. For the differentiable erasing operation proposed, a differentiable search algorithm can be used to search the search space. Experimental results show that the proposed search strategy can significantly improve the performance of multiple classic models on three public datasets.
2. **Regarding the single-branch network architecture in deep learning-based person re-identification**: We proposed a single-branch network architecture

H. Gu et al., *Automated Machine Learning for Person Re-Identification*,
Intelligent Perception and Information Processing,
https://doi.org/10.1007/978-981-95-3433-3_7

based on self-attention—*ASnet*. This single-branch network architecture constructs a multi-scale feature within the designed scale module. It is mainly composed of 1×1 and lightweight 3×3 convolutions. By connecting lightweight 3×3 convolutions, features of the same scale can be obtained, thereby achieving the goal of multi-scale search. To learn different structures and ensure the effectiveness and continuity of modules at different levels, three different regularization loss functions are designed. Experimental results show that the search for the single-branch network architecture can achieve excellent performance with parameters below 2.2M.

3. **Regarding the multi-branch network architecture in deep learning-based person re-identification**: We proposed a multi-scale multi-branch network architecture search for person re-identification—*Auto-ReID+*. This neural network architecture search method is built upon a designed three-branch macro architecture, and the proposed macro architecture search strategy can organically fuse multi-scale, global, and local features. The original differentiable search algorithm introduces an early stopping mechanism, which greatly improves the overall search efficiency and the searched network architecture performance. Experimental results show that the searched multi-branch network architecture can achieve excellent performance with only 5.7M parameters and without ImageNet pre-training.
4. **Regarding the dynamic loss function in deep learning-based person re-identification**: We proposed a dynamic margin Softmax loss function for person re-identification—*LFS-ReID*. This dynamic loss function search method is based on the designed margin Softmax loss function search space. The loss function search space can fully explore the four key dimensions of Softmax loss from the perspectives of gradient, shape, characteristics, and optimization. To dynamically optimize the searched part, a sampling distribution and network weights based on a strengthened cross-layer update two-layer optimization framework are adopted. Among them, the sampling distribution is based on a non-linear truncated Gaussian distribution combined with a specially designed prior distribution. Experimental results show that under the weighting of the dynamic loss function, the performance of several classic models on three public datasets has been effectively improved.
5. **Regarding the static loss function in deep learning-based person re-identification**: We proposed a generalized margin Softmax static loss function for person re-identification—*AutoLoss-GMS*. This static loss function search method is built upon the designed generalized margin Softmax loss function search space. Firstly, the generalized margin Softmax loss function is extracted as two computational graphs and a constant for searching. Then, to further improve search efficiency, a generalized margin Softmax loss rejection rule and equivalent regularization strategy are proposed. This is the first work to search for the loss function's predictor. Experimental results show that the searched loss function can effectively empower several classic network architectures to achieve advanced performance, and these loss functions also exhibit certain direct transferability across datasets and networks.

7.2 Future Work

This book validates the effectiveness of automated machine learning in deep learning-based person re-identification methods from three perspectives: data augmentation, network architecture, and loss functions, achieving satisfactory results. Combining the current development trends of person re-identification, this section presents a vision for further leveraging automated machine learning in this field, summarized as follows:

1. **Person Re-identification Tasks under Other Settings**. The focus of this research is on image-based, closed-world person re-identification tasks, which are the most basic person re-identification tasks. For person re-identification, there are many other tasks under different settings. Current research hotspots include video-based person re-identification tasks (Pan et al. 2022; Yao et al. 2022), cross-modality person re-identification tasks (Lan et al. 2022), occluded person re-identification tasks (Luo et al. 2023; Jia et al. 2023), and unsupervised person re-identification tasks (Wang et al. 2022). Although the methods proposed in this book are validated only on the most basic person re-identification task, they can be applied to person re-identification tasks under other settings after certain adjustments. Taking cross-modality person re-identification tasks as an example, the fundamental difference from basic person re-identification is that in the testing phase, the query image is infrared, while the image gallery contains normal RGB images. In such a setting, the methods proposed in this book can be applied to this task.
2. **Full-Process Automated Machine Learning**. The methods designed in this book perform automated machine learning independently from three aspects. While this is certainly suboptimal, the theoretically optimal approach is a full-process automated search across all three aspects. However, this poses a huge challenge to the entire search algorithm. Currently, some existing works perform joint search for training hyperparameters (e.g., learning rate and weight decay) and network architecture, or joint search for data augmentation and network architecture. However, a full process that also includes loss function search is almost non-existent. Loss function search itself is an extremely challenging task, and considering data augmentation and network architecture alongside it makes this joint optimization very difficult. Nevertheless, this is the ultimate goal pursued by automated machine learning.

References

Xiao Pan, Hao Luo, Wei Jiang, Jianming Zhang, Jianyang Gu, and Peike Li. SFGN: representing the sequence with one super frame for video person re-identification. *Knowl. Based Syst.*, 249:108884, 2022.

Yingmao Yao, Xiaoyan Jiang, Hamido Fujita, and Zhijun Fang. A sparse graph wavelet convolution neural network for video-based person re-identification. *Pattern Recognit.*, 129:108708, 2022.

Xu Lan, Xiatian Zhu, and Shaogang Gong. Unsupervised cross-domain person re-identification by instance and distribution alignment. *Pattern Recognit.*, 124:108514, 2022.

Chuanchen Luo, Chunfeng Song, and Zhaoxiang Zhang. Learning to adapt across dual discrepancy for cross-domain person re-identification. *IEEE Trans. Pattern Anal. Mach. Intell.*, 45(2):1963–1980, 2023.

Mengxi Jia, Xinhua Cheng, Shijian Lu, and Jian Zhang. Learning disentangled representation implicitly via transformer for occluded person re-identification. *IEEE Trans. Multim.*, 25:1294–1305, 2023.

Menglin Wang, Jiachen Li, Baisheng Lai, Xiaojin Gong, and Xian-Sheng Hua. Offline-online associated camera-aware proxies for unsupervised person re-identification. *IEEE Trans. Image Process.*, 31:6548–6561, 2022.

MIX
Papier aus verantwortungsvollen Quellen
Paper from responsible sources
FSC® C105338

If you have any concerns about our products,
you can contact us on
ProductSafety@springernature.com

In case Publisher is established outside the EU,
the EU authorized representative is:
Springer Nature Customer Service Center GmbH
Europaplatz 3, 69115 Heidelberg, Germany

Printed by Libri Plureos GmbH
in Hamburg, Germany